OUT OF THE FIRE
The Loren and Celeste Davis Story

CELESTE DAVIS

xulon
PRESS

Out of the Fire
The Loren and Celeste Davis Story
by Celeste Davis

Printed in the United States of America.

ISBN 9781498474214

Unless otherwise indicated, Scripture quotations taken from the King James Version (KJV) – public domain

Thompson Falls Photo by Justin Gilbert, Nyahururu Kenya 1996
Other photos by Celeste Davis and Loren Davis Ministries with permission

Biography/Autobiography
Inspirational
Christian Missions
Africa
History

www.xulonpress.com

Dedication

But the mercy of the Lord is from everlasting to everlasting,
upon them that fear him, and his righteousness to
children's children.
Psalms 103:17

This book is lovingly dedicated to our parents
Who gave us life and taught us to choose to serve the Lord.
We pass the way to our children and their children.

Table of Contents

Preface

There is an East African tribe which observes a certain wedding tradition. The bridegroom's parents wrap a new blanket around each of the bride's parent's shoulders. These gifts are a show of compassion; understanding that the warmth of the home is leaving them to follow her husband.

God covers us with His love when we leave our old life to follow him. (Authors emphasis)

The story you are about to read is true. My husband and I spent 28 years in Africa and other parts of the world. We never dreamed we would find ourselves in the adventure of a lifetime under the covering of our great and mighty God.

May you find faith and take courage for the trials you are facing in your own walk thru this fragile thing called life.

"and others save with fear, pulling
them out of the fire.." Jude 23

Chapter 1
A New Beginning

We sat in the car watching the planes take off and land. The airport had become a favorite place to stop and talk after church. Loren was a private pilot and loved to fly. West Texas had been a good teacher. It could be beautifully clear or there could be extreme high winds and even snow storms.

I had been an International flight attendant so we had the flying experience in common and the airport had become a comfortable place to unwind and visit.

This day was different. Our tears were falling freely now. Both of us had been deeply hurt by real world 101. Disappointments, disillusionments, rejection and divorce had broken into our lives to try to destroy us. Foolish choices and children needing us to be whole plagued our every thought. These were heavy burdens to share for fear of another rejection but a strong belief in God had brought us together and we decided to talk it out.

We met thru a mutual friend at a Christian radio station where I was working at the time and had my own radio program. Loren had written a novel about end time prophecy and produced it for audio presentation.

The station wanted to air it as a series on Sundays, so he drove a couple of hours over from his town to set that up. When he walked thru the door and saw me he said he felt like Adam must have felt when he had just seen Eve.

Neither one of us had been looking for another relationship but over time we got to know each other as well as anyone could by talking on the phone and seeing each other long distance. He would drive up on Sundays and we would go to church together.

Now, he felt it was time to tell me things that he had locked up in his heart for a long time. He told me he truly thought I would reject him and that would be the end of it, but he felt the relationship had moved to the place where it was time to talk about any future we might have together. He felt he needed to be completely honest about himself and initiated the conversation.

In a surprise to both of us, compassion and tears broke us both as we shared the hurts we had and the fears we each felt. After a long time we decided to get married. I sold my house and we packed up and moved away from everyone and every-thing we knew. It was 1987.

In earlier years Loren had bought a piece of property in the piney woods of East Texas overlooking a lake. He was really an evangelist and would travel out there to preach for small churches in the area. He enjoyed the completely different look from West Texas. When his personal life fell apart he worked as a contractor so he could remain free to take any ministry oppor-tunities that might come along. This dream of fulfilling his call to ministry and hope for some kind of future personal happiness in spite of all that had happened in his life, was by faith what kept him believing God to complete those dreams.

I loved the Lord and had a dream of my own to serve God with all of my heart. I had come to know and understand that Christ was the only stable thing in the world that I could trust. Now, God had brought us together and Loren began to feel that with my strength and love we could do the things God had put in our heart. He was raised in a pastor's family and accepted Christ at a young age. Both his parents were pastors and he was the youth pastor in his dad's church. Soon he was preaching the Sunday night services and was a leader in his college and community. He had gotten a BA with a minor in anthropology and social science knowing he was going to be an evangelist to the nations. All those dreams had been dashed to pieces by the time we met, but now, it felt like God's promise might be possible for both of us.

There had been so many miracles along the way to help him get the foundation up on the house. It was only framed up at this time and we rented a small place to live in while we worked to finish our new home. The sale of my house gave us the confidence to continue to build the house with cash as we had it. Loren was working as a contractor in West Texas but we didn't realize the move would be so traumatic for our children and us. Leaving their friends and also moving from a city that had everything to a small town with limited opportunities turned our dream into a nightmare pretty quickly.

We had arrived with excitement. The plan was to continue his business where he had always been able to hustle and get plenty of work. I was encouraging him to go into the ministry fulltime and do what he was called to do, crusade evangelism, but he thought he had plenty of good reasons to delay this. I had never been in ministry and didn't realize why he was so hesitant. I felt if God called you to do something for him, you should do it, even in the face of adversity. Even so, over the years he took his tape recorder with him as he worked and listened to Bible tapes getting ready for when he could go out and preach again. He said he knew it was coming. In a dark hour of his life in earlier years, the Lord spoke to his spirit and said if he would be faithful to Him, He would use him to his fullest potential. He believed it, but like Joseph in the Bible, foolishly spoke his dream and many mocked him and thought nothing would ever come of it. Genesis 37:23.

Loren would tell me I was the most beautiful woman he had ever seen inside and out. He really made me feel that way too. I was so broken by rejection that I needed that affirmation of

17

love. He said my kindness and gentleness was healing to him and he wanted to be the same thing to me. Soon after we arrived in our new town to finish the home, we realized there wasn't much of a market there for his trade. In fact, he could get little business which troubled both of us. A family and a new house to construct was a big responsibility. There was a point where I tried to take the initiative and actually was offered a job as an International flight attendant again, but with the offer came the mandate to have to move to Europe to get the job. Loren was horrified that I went that far into interviews without his knowledge and I actually thought, like Sarai in the Bible, that this might be God's way of working out our financial problems. Needless to say I had to back off "my plan".

One day he and the kids were cleaning up the property on our new place. They wore shorts as they cut down brush and weeds, picking it up with their hands, to clear the land. At the end of the day they decided to just have some fun. Hopefully a hot game of basketball would work the frustration out of all of them but while playing an aggressive game, a sharp pain hit Loren at the ankle and he slumped to the ground helplessly.

The boys helped get him inside our shell of a house where I propped up his leg and prayed for him. The next morning, to our dismay, he was broken out in poison ivy from the top of his head to the bottom of his feet. Not being raised in the country, he didn't realize when they were cleaning up outside and carrying the weeds away, that it was poison ivy. They had even been burning it. It's a miracle it didn't get into their lungs.

We went to the doctor to do something about the break out, but after examining him, the doctor said the Achilles tendon had snapped and he was more concerned about that than the poison ivy. They told us he needed immediate surgery. After seeing the surgeon, he determined Loren was too contaminated with poison ivy to operate, and instead put him in a full leg cast from the top of his thigh to his foot. He was in bad shape.

We already had a difficult time finding jobs, but now this put us in an impossible situation. We felt led to try to get services to preach since we could at least stand up and talk. Now, we primarily lived on savings and this took away from finishing our new house. It was a torturous situation for a man who needed to take care of his family and frightening to me as a wife and mother.

We decided to rent small halls in different towns in the surrounding area and independently do seminars on exposing Satanism. We had been deceived and were angry at how Satan had tried to destroy us and those we loved. "The thief cometh not but for to steal, and to kill, and to destroy: I am come that they might have life, and that they might have it more abundantly." John 10:10. We wanted people to know who their enemy was and that their best friend was God. This was during a time when Satanists had killed a young college student in a ritualistic killing in Mexico. The papers were full of such stories as they tried to get to the bottom of the riddle of satanic ritual killings. One day, we went to a small town to promote a seminar by advertising with handbills and posters we put up in local businesses. Loren was still pretty bad looking from the poison ivy. One storeowner looked at him on crutches with a full leg cast; eye nearly swollen shut, and lips all swollen; it looked like he had leprosy. The owner took one look at the poster and then looked at us and said, "Sir, if I was you, I think I would leave him [Satan] alone." We all laughed and I said, "If you think he looks bad, you should see what the devil looks like. My husband is the winner!" I was always trying to encourage him no matter what.

Financially, things got so we could no longer afford the rent on the singlewide mobile home we lived in while we struggled to finish our house. This forced us to move into our new place, finished or not. When I say "unfinished", I mean much of the house had no sheetrock; it was wall to wall concrete; and the day we moved in, we put up cabinets, water heaters, and set the toilets.

Loren made the mistake of telling his brother we moved in that weekend. What he didn't tell him was it was out of dire necessity, not because the house was finished. His mother, brother, and his brother's wife, drove in from out of town to see the house and rejoice with us at finally getting moved in. They were waiting in the driveway when we drove up with a load from the rented house. They didn't say a word, but stared at us as though we were out of our minds. They could visibly see we were a long way from finished. Loren was not about to debate about that issue. However, people were amazed at our faith in building this wonderful house without a loan. We had made up our minds to be debt free so we could accomplish the larger goal of ministry. The Lord had impressed on us to cut up our credit cards and get out of debt. He spoke to our spirits that believing and following His word to help us would be a small thing in comparison to what we would believe Him for in the future. We would pour over the Book of Proverbs and other scriptures to glean understanding of how to do this. We wrote down every scripture having to do with money or business and prayed a lot for wisdom and understanding.

I don't know what was worse, the snapped Achilles tendon or the poison ivy. It had covered Loren's leg under the full-length cast. He would take a clothes hanger and poke it in the cast to scratch to try to get a little relief from the itching. One of the most humbling things happened when a woman gave us some food stamps as an offering in one of our services. When you have three kids at home, and two of them play football, you are grateful.

Loren had an older gold Mercedes which was bought while in business. Now we had been reduced so low, he had to take the food stamps to get some groceries, something we thought we would never do. He drove the Mercedes to the side of the grocery store, put on his sunglasses, pulled his hat down to cover much of his face, and went in. It's amazing a store detective didn't profile him and have him arrested as a thug but he had swallowed his pride.

Thanksgiving was coming and we still had somehow survived. He could do little work on the house due to his condition and we had no money for a Thanksgiving dinner. I learned to fix ground hamburger every way you could imagine and had read a wonderful testimony of a Christian family who had nothing to eat and how they prayed and set the table, expecting God to provide for them. I was so encouraged by that story and believed God could do the same for us. During this time of tribulation, the Lord was building up great faith and trust in Him.

One day late in the afternoon, there was a knock on the door. By the time we got there, we couldn't find anyone, but on the doorstep we found three large bags of groceries and a huge turkey with everything you could imagine for a fabulous Thanksgiving meal. Believe me we all had much to be thankful for. Whoever had left the food not only left us a wonderful Thanksgiving meal, but had also included a lot of junk food for the kids. This was a humbling period for us and a testimony that God had remembered the kids with chips and cookies and soda.

I would get up in the mornings and have my quiet time with the Lord in a chair in a far corner on the concrete family room floor, crying and praying. I felt so discouraged for us but, after a few minutes, I would get up and finish my makeup, put on a smile, and be okay for the rest of the day. I didn't realize Loren was watching and it made him feel badly about our situation. I didn't know how to handle it either and would just get into intercession with the Lord, praying for a breakthrough.

Loren had many miracles of people helping him get the house framed up and dried in, but after we moved in, people assumed since the outside of the house was finished we were ok. He wasn't a carpenter and it put us in a fix. He would go to building sites to see how they did things, and come home and emulate the work on the house. We tore out many an interior wall. It's a good thing we were committed or we would have killed each other.

One day, Loren was really proud of himself. He had put up a light panel in the guest bathroom. I had gone out to do some errands and when I came in, he proudly took me in there to show me what he had done. Instead of praising him, I informed him that it was crooked. A big lively debate ensued. He told me he had used a level and that it was straight. I said I didn't care what the level said, I had a perfect eye and it was crooked. We vehemently disagreed but Loren agreed I had a perfect eye to pick him. To break the tension I jumped on his back, remembering a story we had heard about another preacher and his wife, and said, "It's crooked." Loren had a hammer in his hand and jokingly acted like he would hit me to get me off his back. By this time, the kids entered the picture, and as usual, sided with me and told Loren to give up. He straightened the light fixture.

In the middle of all this a Pastor friend from Florida invited us to go down to Haiti with them and preach in a conference. They would pay all our expenses. We scraped up every bit of finances we could and drove to Florida and flew over from there. God moved mightily in those meetings and the last day of a three-day meeting, the amphitheater was packed with over three thousand people. We were beginning to get excited. We had been praying for open doors.

By this time we realized God was actually closing the door to business and Loren was to begin preaching in earnest. For a pastor to open his pulpit to an evangelist he didn't know is almost impossible. We decided to attend a big minister's conference to be encouraged by the messages and to try to make some friends. After the first morning service, Loren went over to where the visiting foreign preachers were and I went browsing through the church bookstore. Walking around the area, he met and began talking to some men from Tanzania East Africa.

As soon as they saw him, he said one of them said, "You're the one."

Loren said, "I'm the one for what?"

The man said, "You are the one we came looking for to preach our conference and crusade in Tanzania." Loren asked him when this was supposed to happen. He answered excitedly, "In six weeks. Give me the answer now and we will wire Tanzania and set things up."

The last thing I had told Loren on our way to the conference was, "Loren, we are so low on funds please don't obligate us for anything." He was a real giver in his heart and would give away everything we had if he could but I had the gift of administration. He was an evangelist and well, sometimes, this caused friction. At this point in fact, we had barely enough to fill our tank with gas to get to the meeting, which was only two hours away from our home, so there was nothing to administrate. The Tanzanian men were so insistent on an answer immediately that Loren told them to wait until he got his wife out of the bookstore. Without a doubt, he believed this would kill the whole thing because I was so frugal. He found me and took me by the hand, practically pulling me out of the bookstore and insisted I come with him.

Sensing something traumatic had happened, but not knowing what, I asked, "Where are we going?"

He said, "Just come with me." I reluctantly followed, almost tripping in my heels, since he gave me no choice.

When we reached the Tanzanian ministers, he told them to tell me what they had proposed. The spokesman said, "We came here to find the man who was to come and preach to our country."

Another one of the men said, "This man is considered a prophet in Africa. You had better listen."

Immediately, I answered, "It sounds like God to me." I looked at it as an answer to our prayers, but Loren stared at me in bewilderment thinking I must be schizophrenic. After a discussion with them, we told them we would spend the night praying about it, but Loren thought, *why pray? We just don't have the money.*

Finally, after stomping around the floor all night, almost shouting at the Lord why it was impossible for us to go, the Lord quieted our spirit and told us this was His will. We told the team from Tanzania we would come to preach in six weeks as they had asked. Mind you, we hardly had money to eat, much less go to Africa. Now, the clock ticked and the countdown had begun. We stayed in contact with two of the visiting African Pastors, in fact, bringing them to stay and preach in our area and introducing them to the local churches.

If it were possible, financially things got even tighter. We thought, "Lord, what are we going to do?"

The Lord spoke to Loren's spirit and said, "Give. You preach my word, give and it shall be given unto you. Luke 6:38 Do you believe the word you've been preaching?" We were challenged. Loren said, "But Lord, we don't have any money to give."

By faith, we had been giving tithes and small offerings but still struggling. Now He was asking us to "prove me" saith Lord of hosts, from Malachi 3:10.

In his spirit Loren said he heard, "Give your computer."

He said, "But Lord, that is about the only thing we have left." In, fact, someone had given him that computer setup and we had never even unpacked it because of all the moving. That was the last thing we heard from God. Now things went silent. Loren really struggled with what the Lord had said to him, but then he felt prompted in his spirit not only to give the computer, but also to give it to another preacher who was seemingly well off. This part annoyed him, but we needed a miracle. He packed the whole thing into the car, put what little money we had toward gas, and drove an hour to this man's house.

To his delight, the man wasn't home. He thought, *Praise God, the devil nearly stole our computer.* Then he felt a quickening of the spirit to leave it anyway. He knew the man personally and even though he wasn't home, he found a way to leave the computer. It is a good thing the police didn't come. We could envision the conversation that would have followed:

"Why are you breaking into this house?"

"Sir, I'm giving this man a computer."

That would have been a hard story to swallow, but after leaving the computer, he said his faith lifted high and felt joy all the way home knowing it was God's move now. He had been obedient.

The pastor of the church we attended said they would buy our plane tickets to Africa and would make sure our bills were covered while we were gone. The church loved us and was enthusiastic to help us all they could. We would be gone for six weeks. Our parents said they would take care of the kids while we were gone and since it was summer vacation that made it more convenient not to have to deal with school schedules.

We sent off our passports to get our visas from the Kenyan and Tanzanian embassies in Washington, D.C. A few days before we were to leave for Africa, we took the calculator and ran the numbers to see where we were financially. The group from Africa had said they would feed and house us once we arrived but we knew we had to have reserves with us. Loren said he thought this must have been a demon-possessed calculator because we were at a minimum of one thousand dollars short of what we thought we would possibly need to live in Tanzania for six weeks. He decided to stop by and talk with another pastor friend of ours. After sharing our problem, they looked at each other with a blank stare with no solution to this seemingly impossible situation. There was no sign of intelligence in either one of their eyes, but after a few moments, the pastor picked up his phone and called someone.

After a short conversation, he hung up and said, "The Lord just blessed you with another $1000."

We began to praise God. "Who was this?" he asked.

The man who gave that donation was a businessman in town. He was a man we had casually met. I had written a short letter

25

detailing our call to preach in Africa and asking for help, if the Lord led anyone, for our trip. I made copies of this and handed them out to everyone we had ever met. The Lord dealt with this man about helping us, but since he hardly knew us, he was in a quandary. He didn't know what to do. He wasn't sure if he heard from God or not. He later told us on the night of the phone call, he was reading in the book of Proverbs about casting lots so he took a coin out of his pocket and flipped it. Heads, the Davis's get one thousand dollars; tails, nothing. It landed on heads. The angels had to have gotten involved in this. He had flipped the coin about the time the pastor had called him and was ready to give the one thousand dollars we needed. Now we had the bare minimum to take with us, but had not received our passports back from the embassies yet.

Weeks earlier, I had called the airlines to find out how much the tickets were and to get details of when we would have to pay for them. The reservations agent told us that we needed to make up our minds how we wanted to travel. Of course, it was economy, but she also gave us an option of routes. We could fly to Europe and change planes to fly directly to Dar es Salaam Tanzania or go through Europe and change planes to route thru Nairobi, Kenya, the country north of Tanzania. Both of us had traveled extensively, but neither one of us had traveled to Africa before and had no idea of the importance of the decision we were about to make. We had prayed for the Lord's guidance in every single matter concerning this trip and now when this question came, I was *impressed* by the Holy Spirit to get our tickets by way of Nairobi Kenya. For this reason, we sent our passports to both the Kenyan and Tanzanian Embassies in Washington, D.C. to get visas. We would have to change planes in Nairobi for our onward journey to Dar es Salaam in Tanzania, but were able to have a layover time in Kenya if we wanted.

It was two days before our scheduled flight and the passports still hadn't come. That morning Loren called the Tanzanian consulate trying to find out where our passports were. The lady answering the call was nonchalant and said she was sorry, but could not locate our visas and perhaps they had lost our passports. Panic stricken, he almost yelled at her, saying we were scheduled to leave the country in two days. Alarmed, we rushed over to visit with an elderly prayer partner, and told her about this situation. This precious saint was a real prayer warrior. She reminded us of the power of prayer in the name of Jesus. Matthew 16:19. We held hands and began to pray. She commanded the devil to turn our passports loose and asked the angels of God to retrieve them from wherever they were and get them to us right away. We went home encouraged and knew that she had touched heaven.

Later in the day we called the Kenyan consulate in Washington, D.C. to see if our passports might be there. After searching for a few moments, the lady there said, "Brother, I have good news and bad news. The good news is I have your passports in my hand. The bad news is Tanzania had given you visas, but when they realized you were coming in as missionaries, they cancelled them."

At this time, Tanzania had an Islamic government. When they found out we would arrive there to preach the gospel, they cancelled our visas. When you apply for a visa anywhere you are asked on the form to state the reason you are coming into the country. We simply told the truth about coming in to preach. No one warned us of the implications of that truthful statement. We were horrified, but the woman in the Kenya embassy in Washington, D.C. said, "Do not be dismayed, I'm going to give you your visas to Kenya. Go in faith and God will open the door to Tanzania for you like He opened the prison doors for Paul and Silas." Acts 16:25-26.

God had placed this total stranger from the embassy into our life and she was a Christian. We were so excited. We began to see the confirmation by the hand of God that He truly was leading us. She sent our passports by overnight mail and, although we felt like we were in a corner, we decided to take a step of faith and go to Africa not knowing anyone in Kenya or how to get into Tanzania. Had I not been able to discern the voice of the Spirit directing us to go through Kenya weeks before, we would not have been able to leave the United States with cancelled visas to Tanzania. Excitedly, the next day, we boarded the plane and headed for Africa.

Chapter 2

Lions and Snakes

Green is definitely the word for what we were. We knew absolutely nothing about Africa and never dreamed of the treachery that might be before us there. Our plane landed at midnight in Nairobi two days later. We had no booking for a hotel and after arriving we found out it was the high season for tourists going on safaris to the big game parks so everything was booked up.

When we got out of customs, there was a man waving at us. He said, "Are you Davis?"

Loren said, "Yes."

The man said, "The Lord woke me up and showed me your face and said *Davis was coming* and needed help at the airport. I'm here to help you."

It was incredible. This man said the Lord, in the middle of the night, had wakened him and He had directed him to come to us. He helped us to get to the hotel a pastor friend of ours had told us about. Without a reservation, in the middle of the night, in a city where all hotels were full, this one had one room vacant with a single bed. We took it and slept peacefully scrunched up on a twin-sized bed realizing God was in control of this trip more than we were. We never saw or heard from that man the

Lord sent to us ever again. The next morning we began to try to get into Tanzania.

Meanwhile, we were not to know until a week later that the men who had invited us from the conference back in the states and who were to meet us in Dar es Salaam, had their own problems. They had gotten as far as London, but their papers weren't right, so London sent them back to the U.S. Then the U.S. sent them back to London. This scenario happened to them twice. They were having their own nightmare. Now, the timetable was completely off for everyone.

To compound the confusion, the hotel manager said our room had to be vacated because it was reserved. The hotel was overbooked and since we had no reservation, we would have to find another place to stay. We told him we had no place to go. He simply apologized. However, we were able to leave our things in the baggage room. As we were walking out of that hotel we bumped into a man who told us about a missionary guesthouse nearby. We immediately headed there.

It was nine a.m. and we had little money left in our pockets to stay six weeks, and had not even gotten to our final destination. Furthermore, our plane tickets were non-changeable and non-refundable. We bought the cheapest tickets we could buy, but the downside was not only that we couldn't go to Tanzania, but we also didn't have the money to stay in Kenya. Finally, we were able to get a promise of a room for one night. The missionary Guest House was actually fully booked, but it was still morning and one room had not checked in. We were told if those missionaries did not check in by six p.m., we could have the room.

It was Saturday and we had only been gone from home a few days and under the stress of no place to stay and limited money, we decided to use the time of waiting to go to the Tanzanian Embassy and try to get our visas. After sharing our plight with a missionary at the guesthouse, he tried to comfort us. He said, "I have been here for twenty years and I have never known anyone to ever get a visa to Tanzania once they had one cancelled." He would have competed with Job's comforters. Job 32:3.

Loren said, "We have to try. There's no choice. God does miracles for people who believe in His word. He had spoken to us to come here so it is His business to make a way."

I gave him the passports and we walked the dusty road to the Tanzanian embassy in Nairobi town. Raking up all the power of positive thinking we could muster, we told the lady we needed visas. She saw the cancelled visas in our passports and glared at us. We smiled anyway and she passed us the request sheets. We were told on Saturday everything closes at noon. We went ahead and filled out the papers. The instructions said we needed some passport photos and photocopies of our tickets and it was almost eleven a.m. We decided to hit the street looking for a photo shop and a copy machine. We quickly got our photos taken and while I waited for them to be developed, Loren wandered about looking for a copy machine. He found himself in an old building with a sign that said "Photocopies Upstairs".

A man met him at the top of the stairs and said, "You need help with your passports, don't you?"

Surprised, Loren answered, "Come to think of it, I do."

He shared our problem and the man said, "I can help you. I know someone at the Tanzanian Embassy. Give me your passports."

Loren gave them to him and left praising the Lord. When he told me what had happened, I said, "You did what? You have ruined us! You never give anyone your passport. Stolen passports are hot items."

I was so distraught and now we were both very nervous at this possibility. Loren had the gift of faith about everything and it was one of the qualities that drew me to him. I often called him my "absent minded professor" because we didn't think alike sometimes. I was much more cautious and we counterbalanced each other. We were stunned at this possibility, but we had nothing to do now but go back to the guest house and see if we had a place to sleep. Maybe this is why the scriptures say, "Not many wise ... are called". 1 Corinthians 1:26-27.

God moved for us again when the other missionaries did not check in. We paid six dollars for our room and tried to sleep, wondering about our passports, but believed God had done too much to get us this far to fail. When we went back on Monday to where Loren had met that man in his quaint office, incredibly he had our passports with visas to Tanzania. I shook my head in wonderment and we both praised God for his hand of protection.

We flew the next day into Dar es Salaam, Tanzania several days off our scheduled time. When we arrived while waiting in line for immigration, we read on our entry papers we had to show proof of fifty dollars a day per person for the duration of our stay in Tanzania and proof of a return ticket. We had the return ticket but the funds again were another big problem. Loren hadn't told me exactly how much we had and I trusted him completely and never asked. Now, I found out we did not have even barely enough to stay the six more weeks. We began to pray in the Spirit under our breath, we were next in line to speak to the agents.

We knew we needed another miracle. About that time we heard our name called out, "Brother Davis, Brother Davis," Surprisingly, it was one of the men we had met in Houston, but it was the one man out of the entire group who didn't want us to come. Why he was there is inexplicable. Somehow he had managed to pass through customs from the other side and came up to a man in the immigration booth, a man he obviously knew. He came over and pulled us toward his friend and said, referring to us, "This is a big man and his wife from America. Please, let them come through quickly."

The immigration agent let us through without checking our finances and barely looked at our papers. God had made a way in the wilderness. Isaiah 43:19. To our surprise and delight, there was a big delegation of pastors outside the airport baggage area waiting for us. How they knew we had come in on that flight, we will never know since we were so off schedule. We didn't know we would arrive ourselves until the night before.

The other original contacts from Tanzania we had met in America still hadn't arrived from the seesawing across the ocean. We didn't know where they were or even that they had problems. The group who met us at the airport was so kind and after spending a night in Dar, put us on a train with four other local Tanzanian pastors and we began a two day trek across the interior of Tanzania to the southernmost town of Mbeya. We felt like we were in a movie but the romantic colonial train from the movies was, in fact, a cattle car for people. Our "first class" compartment consisted of two wooden benches on each side of a six-foot area, sleeping bunks on top, and a sliding door. "First class" meant we didn't have to share our room with anyone else, however, the four pastors with us made themselves at home. We came to learn "first class" and "private" meant two different things. The air conditioner was an open window and the public toilet located down the hall had an unbearable stench. There was no holding tank; only an open hole in the floor. The train would stop at each and every village at least every half hour nearly knocking you off the wooden bench. It was so hot inside and outside and we were warned we couldn't walk around the train or open the windows very wide because thugs would try to jump in at the stops.

On this trip, we stared at each other a lot; no words were necessary. We each knew the other's thoughts, yet we still had great hope and fire in our hearts. One of the most memorable moments of our life, in spite of the mode of travel, was rounding a curve and hearing the train screech to a slow halt. We stood up and looked out the window to see why we had unexpectedly stopped and saw three majestic lions laying on the tracks just ahead of us. Now for sure, we knew we were in Africa. It wasn't a dream.

On the way to Mbeya, we tried to get some information about the schedule they had planned for us. Loren asked our escorts when the crusade would be held. Our host said, "What crusade?"

Startled, Loren said, "What do you mean, 'what crusade?' That's why we came."

They said, "Crusades have been outlawed in Tanzania. You are preaching a conference."

Loren was visibly upset and said, "You mean my wife and I have been through all this stress to get here and there is no crusade? Don't tell us that."

Our host answered, "I'm sorry, but that is the truth." We were so distraught. Suddenly we felt lied to and used. We had no words. We didn't understand then, but The Lord was stretching us. He was making us "meet for the master's use". 2 Timothy 2:21. We pondered and prayed.

Finally, after two grueling, long hot days and nights on the train, we expressed a big sigh of relief when we finally arrived at Mbeya. It was cold and a pleasant change from the heat we had experienced in Dar es Salaam and throughout our long journey. Now, we were excited again. Loren began preaching in the youth conference as planned and God moved on the young people. In Africa, "youth" means from fifteen to thirty-five years old. We thought we had started late in life due to all our personal troubles but now realized we weren't so old after all.

The next morning, was to be my time scheduled to speak, but after getting up early and ready to go, no one came for us. After a few hours of stressful waiting and again wondering what we had done in coming here, someone finally came to escort us. By this time, Loren was offended because it seemed so discourteous that we were treated this way with no communication. I was wrestling with thoughts of a "woman preacher thing" and my own insecurities in this new culture. After all we had gone through to get there, and to be treated so casually and with such seeming disrespect, infuriated Loren. They apologized and we forgave them, of course, but later on we began to realize that this was just our first encounter with *"African time."* It was six hours later than the schedule we had been given; but God gave me the grace and peace to teach His word calmly and with His strength in my weakness. We began to realize that our American ways, emotions and expectations were being challenged to the max by this experience and there was no choice but to let God change *us*. We were on the potter's wheel and He was crushing the clay. Jeremiah 18: 2-4.

The biggest concern was that no crusade was scheduled and we had told the few people who had partnered with us that we were going to Africa to preach a crusade as we had been told. Loren began to push for a crusade. Our hosts told us it was impossible because of their Muslim government, and the church was fragile and in danger of sinking. Loren told them, "It's too late. If you are not evangelizing, it's already sunk." There was nothing left to lose. Our thoughts were that we were not willing to see the millions of people in Tanzania unreached for Christ while the church hunkered down shaking in fear.

SUMBAWANGA

A preacher was at the conference in Mbeya from a small town called Sumbawanga deep in the western interior of Tanzania. He heard Loren preaching and of our desire to hold a crusade. He asked if we would come to Sumbawanga and preach a crusade in his town. Without hesitation we said yes, having no idea where Sumbawanga was or what it was like. All we knew was if we didn't hold a crusade, we were finished at home before we got started. We liked his zeal to get people saved; we had a kindred spirit.

The Pastor hired a small plane to fly us to Sumbawanga. This was a big step of faith for him and a positive indicator to us of how serious he was about soul winning. The night before we left, he sat down with us and told us a little about Sumbawanga. He said *Sumbawanga* means, "to throw witchcraft". The Pastor said this place had the most powerful witchdoctors in East Africa and said people came from all over Africa to understand how to put powerful curses on their enemies. We thought to ourselves, *we must be desperate to preach a crusade in a place like this*. We had prepared to fly out the next day but the night before we had a demonic visitation in our room. It was a horrible presence, dark and heavy. It was like a deep guttural voice threatening, "If you go to my city, I will destroy you."

We grabbed hands in the darkness and said, "Do you sense that?" Each one of us, unknown to the other, had felt this evil presence. We held hands, prayed aloud, and the presence left and we went to sleep, but not before telling the devil, "You have tried to destroy us for a long time. If we go home, we will still have to fight you. So if we're going to have to fight, we're fighting you or your own turf. We're coming to the fight."

The next morning we got into a small, single-engine plane with the Pastor and flew out of Mbeya to only God and the pastor knew where. We found ourselves flying over croc filled swamps and deep jungle; believe me, we prayed for that little single engine plane. We couldn't take our eyes off the propeller to make sure it still turned.

Two hours later, the plane finally circled a small village with a dirt airstrip and then we landed in Sumbawanga. A few natives gathered around the plane to see the *Wazungu* "white people; or, technically "Europeans". I was excited and happy to be on the ground. When Loren got out of the plane, he put his hands on his hips like John Wayne and said, "Devil, I'm here." I don't think we needed to inform him. The air was thick with oppression.

True to his word, the Pastor went to the local government authorities and secured a crusade permit for three days. Sumbawanga didn't have much lumber available, but we found enough used wood to build a crude platform about one foot high, and four foot by six foot wide. We had two borrowed horn speakers and a borrowed small amplifier. We put a crowd control rope around the platform in faith there would be a crowd.

That first night of the crusade in Sumbawanga is one we will not forget as long as we live. We had about a dozen Christians from the church. About five hundred people showed up, mostly Muslims, pagans, and witchdoctors. It felt like many were demon-possessed in one way or another. This Pastor was bold in bringing us to Sumbawanga, but we noticed no one sang except the small but brave handful of Christians over to the right side of the platform, and they were very timid like they didn't want the devil to notice them and get mad. We had no idea what we had gotten into.

Loren began preaching, "I have come today to break the curse that is over your life. The Bible says, 'cursed is every man who hangs on a tree.' Two thousand years ago God's only begotten son, Jesus Christ, hung on a tree, the old rugged cross. Everyone who looks to Jesus will have the curse broken off his life. John 3:16-18.

The Muslims paced back and forth in the back of the field and much of the crowd jeered and mocked. I was on the ground, standing to the right of the small platform, videotaping with a borrowed camera from our dear friends Marge and Jerel Bower back home. I wanted to document the first day to preach a crusade in Africa. This was my first experience in a crusade too, although I had recognized Loren's preaching gift in a church setting and could see he had a call of God on his life.

Down in the front row, sitting on the ground by the crowd control rope, was a man laughing menacingly. While Loren preached the Word of God, the man became hideously contorted and was mocking. This began to annoy me. Not that he mocked Loren but that he mocked the Word of God. Suddenly I put the video camera down and walk toward that man with my finger pointed to him. I found myself speaking in tongues according to Mark 16:17 and commanded the mocking spirits to come out of that man. I was from a Baptist background and had never cast out devils before in my life, but had read in the Bible that Jesus said, "In my name you shall cast out devils."

I believed the Bible. Before leaving America I had gone thru a time where I was tormented by fear, actually, I didn't realize it but my whole life was fear based. I was almost convinced I shouldn't go. "After all, I thought, I'm not the preacher. What will I do if I have to pray for someone? Loren is the one with experience and training". He was always sharing stories about his early years in ministry with his oldest brother who was also an evangelist. He had some stories that were thrilling about how the Lord worked, but I had not been exposed to that kind of life. That same night I had gone to sleep thinking on these things, but the Lord had awakened me over this issue. I felt like an over-filled balloon; the Lord impressed my spirit, This is the *power that is going to save people*. "It is by my spirit saith the Lord." Zechariah 4:6. I believed God's Word. Now here was this man, mocking. I didn't pre-meditate doing what I did, but more like I found myself being moved by the spirit, commanding the devils to come out of that man and, at first, nothing seemed to happen. I didn't give up. I kept commanding in the name of Jesus. It was then the Lord showed me a truth. When you speak the Word of God and nothing appears to happen, that is where most Christians would walk away and say the Word of God doesn't work. So, taking God at His word, in faith, I kept commanding those demons to come out. Finally, the man went limp and seemed to calm down. After that, as I turned around to walk back toward the platform to pick up the camera a big commotion ensued. Meanwhile, Loren is preaching thru all this.

I barely took three or four steps away and the crowd around that man screamed and parted like the Red Sea. As I turned to see what the matter was, a black snake, five feet long and as big as a man's wrist, slithered out from where the man was seated. The snake came directly toward me. Men picked up sticks and hit it, trying to kill it or divert its path away from me. With all the screaming and shouting going on, at the same time, Loren was preaching, "Christ has redeemed you from the curse," Galatians 3:13. Muslims were shaking their fists, yelling and mocking. Men were trying to beat the snake but I stood my ground and had no fear. The Lord had let me know in my spirit this was simply "another manifestation of the devil." In front of everyone, while the men were trying to kill the snake, it disappeared, just vanished into thin air. I felt, by the spirit of God, that if I had feared what I saw with my eyes and not trusted in the Word of God, that snake would have struck me down and I would not be here today to tell the story.

This was our first encounter with what Africans call *Jinns*. In the Muslim world, a *Jinn* is a supernatural manifestation of a demon, which can take on either human or animal form. These *Jinns* are used to serve the purposes of their masters or witchdoctors. It is believed they can hurt or even kill people. In our case, it must have been sent on assignment to cause fear and death to stop us from preaching the Gospel. It was an incredible visual testimony of the power of God being greater than the power of the devil. Luke 9:43.

That night a man whom we were told had been "bewitched" (a curse put on him) came to the crusade. He looked like he was pregnant; his stomach was so large. God miraculously delivered and healed him of that oppression. He instantly became thin. It was a marvel. Demons manifested in people and they began writhing like snakes all over the ground, but the small team of Christians we had, all worked together and commanded them to be loosed in the name of Jesus and the people were free and made whole.

After the service, the children of that place were all so happy and excited. They all gathered around us and escorted us through the village streets to the place where we stayed. We led the parade and I sang, "God's got an army marching through the land, deliverance is their song; there's healing in their hands, everlasting joy and gladness in their hearts. In this army I've got a part." This was a popular Christian chorus and I loved to sing and the children loved it and marched with us trying to sing along. They were so excited to see and witness the wonderful works of God. Normally, in the African bush, children are hushed but that day the whole village saw the joy of what had just happened. Everyone was exited.

We went to our hotel which we called the "Sumbawanga Hilton". It was a native African hotel, a cinder block compound in the eastern tradition where the center of activity is the inner courtyard used for cooking and hanging clothes out to dry and just sitting around visiting. The sun finally went down, so we retired for the day and went to sleep at around nine p.m. We were tired but happy at the success of the meeting.

Then it started. We had just closed our eyes when we heard loud sounds of drums and music coming from the middle of the compound. Our eyes popped wide open. It was so loud, after a while, some of the plaster literally fell from the ceiling. With each drumbeat, it felt like demons were trying to come into the room. It sounded and felt like the earth had opened up and all the fires of hell were trying to lap us up to devour us. We turned on our Bible cassette tapes, but the noise was so loud, we couldn't hear them. We prayed but the drums and music were so loud, we could hardly pray.

We were deep in the interior of Tanzania, East Africa, fifteen hundred kilometers from Dar es Salaam and the only white people in the midst of Muslims and witchdoctors. Finally, at eleven thirty p.m., Loren threw off the sheet and got up and put on his pants.

I said, "What are you doing?"

Loren said, "I feel like Popeye, 'I stands all I can stand and I can stands no more.' I'm going out there and shut them down." He had a habit of hiding stress with humor.

I said, "Oh, no you're not. It's dangerous out there."

He ignored my concerns and told me to "lock the door; I'm not putting up with this racket anymore." I didn't argue about locking the door and I started to pray, thankful for his courage. I did feel secure in the Lord and that He was with us. Loren told me as he walked down that dark hallway, unarmed and heading for the center of the compound where they were gathered that he prayed in the Spirit and felt like a lion, although he didn't know what he would face.

As he turned the corner, he saw a large group of people whipped up in frenzy. Witch doctors had come and were trying to put a curse on us. Loren shouted above the noise, "What's going on? I can't sleep." They couldn't believe it. They tried to put a curse on us and he complained we couldn't sleep. I guess they didn't know losing sleep could upset him *that* much.

A few of them could speak bits of English, and told him, "It's dangerous."

He said, "You've got that right. You'd better get out of here now."

One of the men said, "Give us thirty more minutes."

In his tired state Loren said, "Okay."

He came back to the room and the clock finally hit midnight, but they had not slowed down.

Loren said, "They lied to me," and he got up out of the bed again.

I said, "What are you doing now?"

He said, "I'm going out there; and this time I'm going to shut it down. They are going to leave. The Holy Spirit is with us." There was not one ounce of fear in him. Although he was physically alone and they were many, I knew Jesus was with him. When he got to the corner, he put his hands on his hips and shouted, "It's over; shut it down."

One of them looked at him and said, "You're God."

He answered, "I'm not God; I am one of His servants." But we believe because he had on the whole armor of God, the devil didn't know it was Loren inside of it. They saw the covering of God. Ephesians 6:11-12.

Standing to one side, he ordered them out. One by one, they passed by him in the middle of the night but with their heads bowed in shame. One man looked up and said, "I know I need God." That same night, some men were caught tearing the stones off the Pastor's house and tried to get in to attack him, but they also failed.

The last day, half the population of Sumbawanga attended the meeting. The Muslim owner of the hotel where we stayed pleaded with us to stay. He said, "We are now starting to understand about Jesus." The Sumbawanga crusade lasted only three days, but in the spirit we believe it had lasting effects. In fact, years later, we would meet people in other parts of Africa who told us they attended that crusade and were saved. One man, eight years later, said he was there the night the snake manifested and disappeared. He said he saw it with his own eyes. Many miracles happened and so many came to trust the Lord.

The Pastor flew us back to Dar es Salaam, and our original host put us in one of his houses to stay for our last two weeks until time to leave the country. Local Pastors visited us for fellowship and one on one time, but we had no other ministry opportunities. The denomination that had invited us was terrified of the government and chose not to participate in any other meetings even when we were finally given a permit to hold a crusade in Dar es Salaam. The pastors were threatened by their own bishops and told them not to help us or they would lose their churches. The bishops wanted to remain status quo with the government and not cause any waves. When we finally got out of Tanzania, Loren said, "I'm through with Tanzania. I'm never coming back to this place." It was frustrating not being able to preach the Gospel in the open after all we had gone through to get here. We were just beginning to understand the fight to keep people from hearing the salvation message but had not expected it from within the church.

I knew we would come back. Six weeks before, when we descended into Nairobi airspace, the Lord had given me a vision of His Hand holding Africa and saying quietly, "I'm giving you this land." I did not share this with Loren for years, knowing how he felt about this first trip. Like Mary, I just kept these things in my heart. Luke 1:30.

The flight back to the states was with mixed feelings, trying to take in all that had happened. We had been very deeply stirred by our experience. Two long days later we arrived well after midnight. To our surprise, even though we informed our people when we were coming in, no one was at the airport to pick us up. After calling and reminding the pastor we had arrived, he sent someone and another two hours later we were finally picked up. We didn't feel like conquering heroes.

When we got home we discovered our car had been totaled; our boat had sunk in a freak storm on the lake; none of our home bills had been paid even though the church had promised to do so; the woman holding the note on our property threatened to foreclose and even take our house although we had check stubs proving we were up to date on our land payments. While we were in Africa, Loren's brother had preached for our church and they had taken a special offering for us, which came to one thousand dollars. We never saw a dime of it. We were extremely disappointed in the lack of integrity of our pastor. The good thing is, miraculously we survived. As God had taken care of us in Africa, He began to prove He would take care of us in America. We still had a few churches where we could speak and God was faithful. We didn't fully realize the extent to which God was changing us but our faith and vision was growing. This was the beginning of our full time ministry.

Chapter 3
Gorillas

In 1989 we didn't return to Africa, but returned to Port au Prince Haiti for another crusade. One of the churches we preached in wanted to come with us and bring a group of prayer intercessors. These ladies promised they would carry one of the speakers we would use in the crusade with them as a courtesy since they had extra luggage allowance. When they arrived, to our disappointment they didn't have the speaker. They had seen it before we left and had agreed to bring it, so this put us in a bad fix regarding our sound system. One of them rudely told us, "If you prayed more, instead of worrying about all this equipment, you would see God do something." We were having trouble with such ignorance. After all is said and done and prayed over, Evangelism is also a four-letter word: *work*.

A local pastor whom we had met on our first trip sponsored the crusade, but in the middle of the meeting, another visiting preacher from America came and took him off to the other side of the island to help him. This put us in a bad situation because this local pastor was also our interpreter. We quickly searched and found a young man who could interpret from English to Creole for Loren in the crusade. Haitian Creole is a mixture of French and West African languages that emerged after the Atlantic slave trade. Now, our situation was becoming

44

complicated. The pastor had abandoned us and it hurt us that he had given his word and then broken it so easily. We were on our own to fend for ourselves. The women who said they came to be our intercessors in prayer seemed hostile now. They were probably a bit scared but they reacted by putting a bench right in front of the standing crowd and sat, skeptically glaring with their arms folded. All kinds of spirits were loose. After the service that night, we asked them to move from sitting directly in front of the people. We were learning that the gates of hell are not always outside the church but God would still prevail. "... the gates of hell shall not prevail..."Matthew 16:18.

The Lord did miracles like you read about in the Gospels. One night many deaf received their hearing. It was so glorious. Another man came through the crowd and lifted up his crutches, tried to walk, but fell to the ground. Watching, we gasped as he got up and kept trying. After the third fall he got up and walked well. An old man who had been blind for sixty years received his sight. He was so happy. The miracles that happened there were so wonderful and the spirit of God began to deal with our spirit through a scripture. We needed to "make full proof of our ministry". 2 Timothy 4:5. I prayed, "How Lord?" The answer was a video camera. I remembered the borrowed one for Africa. That began a new era for us, as we began to stretch our faith and believe God for the cameras to record all that God would do in this ministry. I had started college on a partial scholarship in Journalism and that training kicked in and all I could think about was "eye witness" news about what the Lord was doing. The crowd was very large and many responded to accept Christ as their savior. We left Haiti with a new vision for equipment and a lot more understanding of the tactics of the enemy.

OUR FIRST MAJOR AFRICAN CRUSADE

Shortly after this Haiti meeting we received correspondence from a big crusade committee in Tanzania to come back and preach in the city of Tanga. The country's policy had now

changed allowing Christian crusades. We were told our breakthrough crusade in Sumbawanga is what triggered this change.
One of the men who originally worked with us took courage
from our love for souls and approached the government to be
fair and allow Christians to conduct open-air meetings.

Tanga is an eighty-nine percent Islamic city on the coast of
the Indian Ocean. While we waited for the crusade to be physically set up by the work crew, we talked some of the African
host pastor's team into going fishing. Loren thought he hit an
especially good bargain by renting a boat and crew for fifty
dollars. When the boat arrived at the pickup point, we realized
why it was such a bargain. It was a small African fishing boat
with nets piled high in the center. Our African team, all dressed
up in suits, had never been out on the ocean since they were all
from inland and they decided they weren't going to go. They
were terrified of the sea and couldn't swim. I was also privately
reluctant, but decided to go for the men's sake. The men, to
save their masculine pride and not be shamed by a woman, got
on board. To help keep everyone calm, I began to sing simple
Christian choruses and the team joined in and took courage.
African Christians love to sing and being on the boat made us
think of all the times Jesus spent on one with the disciples.

Everyone was nervous about the trip. For one thing, the captain was a Muslim and there had been some recent International
incidents. I didn't want the captain to get us out in the middle of
the ocean and for some reason not bring us back. Of course, all
this was in the undercurrent, nothing spoken audibly. We had to
sit on top of the fishing nets. There were no fishing poles so we
used hand held drop lines. What got my attention though was
when the small crew bailed water out of the boat, which they
did the whole trip. When we were some distance out to sea we
realized there were no life jackets and were unable to see land
anymore. We stopped on a small sand bar island to look for
shells. There was no easy way to get off the boat, so we had to
jump in and swim a few feet. The Pastors in their suits had long

since shed their shoes and sox and jackets. They rolled up their pants and got in too. The captain jumped in after us and pulled the boat close to the sand bar with a rope. We enjoyed the stop, but what struck me was that we were *way* out there. There was nothing but ocean in every direction and it was a long way back. I felt like we were specks of salt and pepper in a bowl of soup. Getting back into the boat, the captain got stung in the chest by a large jellyfish. It was so painful to watch him suffer, but he recovered without a complaint. We caught a few fish, but the most important thing about the trip was that we made it back.

The Tanga crusade was incredible. Loren told the people that Jesus Christ, the son of God, would save and heal them. The first night so many crippled and diseased people came. Sitting on the platform, I could see Loren very seriously estimating the crowd. Afterward he told me what he was thinking as he looked out at the people waiting expectantly. *"What am I doing here? How can I promise miracles for these incurables?"*

Then the Lord spoke to his spirit while his knees knocked in fear. "You are not the miracle worker, I am. Your job is to preach the Word; my job is to do the saving and the healing. You do your job and I will do my job. I am watching you twenty-four hours a day. I'm watching how you treat your wife and everyone around you. I work with people who love and obey me. Turning a city to me and having miracles in your ministry do not come from a formula. They come from a relationship with me and I love them more than you do".

He gave a sigh of relief, and began to boldly preach the promises of God's Word.

Loren held up the Bible in this Muslim city and said, "I hold in my hand the last Testament of God. Jesus is a prophet, but He is more than a prophet. He is the Son of God, and I am going to prove it. When I pray in Jesus' name, demons are going to come out and miracles of salvation and healing are going to happen." God loves all people. It is not His will that any should perish but that all should come to repentance. 2 Peter 3:9.

47

"And they went forth, and preached everywhere, the Lord working with them, and confirming His word with signs following." Mark 16:20.

Later we were told Muslim men were in the back of the field shaking their fists and shouting, "It's not true. Nothing is going to happen."

As Loren began to preach about the saving and delivering power of Jesus Christ, the demon-possessed screamed and fell, writhing on the ground all over the field. We had never seen anything like this, except in our small crusade in Sumbawanga.

Our workers picked them up and carried them to what we called our intensive care unit tents where we had teams trained and ready to cast out devils. It looked and sounded like a battlefield with all the screaming going on, but Loren kept preaching Jesus. Those Muslim men who yelled and said nothing was going to happen were suddenly knocked to the ground by the power of God and ended up in our ICU tents packed with people being delivered.

Great miracles happened all over the field. Cripples and others who were sick had been brought in truckloads and many were healed, and walked across the platform praising God and sharing their testimonies. It was sheer bedlam. Many Muslims were openly coming to Christ, even elders. One Muslim man told me people all over Tanga were saying, "Jesus has come to Tanga." Without a doubt, He had. A Hindu woman with the red Bindi mark on her forehead came forward and accepted Christ and was mightily baptized in the Holy Ghost. One of the organizers of the crusade told me, "The Muslims are shaken. They saw a man they knew who was crippled and now he is walking. The whole city is shaken. The sheiks are shaken."

A LUTHERAN CHURCH

We left Tanga and returned to Dar es Salaam and preached a mini open-air crusade in a Lutheran compound. After the first service, the team took us to the church office, asking us to pray for a demon-possessed woman. She couldn't speak English, but when she saw Loren, she spoke in perfect English, saying, "I know who you are. You just came from Haiti." We had told no one in Africa we had been in Haiti, but we came to the sudden realization that in the spirit realm it was evident the demons in Haiti had communicated with the demons in Tanzania.

Loren said, "That's right, and I am going to do as I did in Haiti. He began to command "Come out of her, in the name of Jesus." She was completely delivered into her right mind. Mark 16:15.

One night during this meeting Loren was unable to preach because he had a high fever and was sick with malaria. He sat on the platform and I took the service that night. It was a mighty message of salvation and deliverance and by the time the altar call came, Loren was healed. That night a man testified God healed him of blindness.

As the meeting continued, an albino man who was a drunkard walked by the church headed for a bar. The Spirit of God drew him into the crusade and he was gloriously saved. He removed a fat brown pouch from around his neck, which he had obtained from a witchdoctor and threw it away, then he came to the platform and insisted on praising the Lord with his homemade flute. It was absolutely beautiful. The sound of his flute was a glorious song of redemption. A woman whose arm had been broken and couldn't move it, testified while standing in the crowd listening to the story of Jesus, she was totally healed and waved her arm everywhere. Another man who was a total drunkard was delivered and saved. Acts 10:38 says "He (Jesus) went about doing good and healing all that were oppressed of the devil, for God was with him."

HONDURAS

Back home in the states, we met a missionary to Honduras in Central America. He invited us to come to Tegucigalpa for a crusade. Being hungry to preach the Word anywhere, we accepted. We had a hard time finding a field for the crusade but Loren showed our host a field between two bridges we had seen downtown. It would be a perfect venue. Although the location was strategic, it was actually a repulsive place because it was a field of human dung. The area was full of drug addicts and hopeless people laying around in a stupor sniffing glue. We told our host this is where Jesus would be if He were here. Loren insisted, and against the missionary's better judgment, he agreed with him to use the field. We had a grader come in and clear the dung off the field and set the platform up by that polluted river. Our host almost balked at doing this, the smell was so bad, but managed to pull himself together and did it anyway. One evening he told us he was so discouraged while working to prepare the grounds. Suddenly a white bird came and hovered over the platform for some time. After this, he stopped complaining and finished his task, taking this as a sign that God truly would do something here if we were willing to prepare for it.

Oh, how God showed up in this old riverbed. Before the crusade, we went on the radio live for several days, building people's faith to come and be saved and receive their healing from the Lord. One day as we were leaving the radio studio, a woman stood in the hallway not letting Loren pass. She said, "I was in a village and heard you on the radio telling about God's healing power. I have leprosy. Heal me" she demanded.

We were taken aback. Loren said, "I will pray for you when I preach the word of God and pray for everyone at the crusade."

She said, "No, pray for me now." We realized this woman would not take no for an answer, so Loren led her to the Lord right there; laid hands on her, and rebuked the leprosy in Jesus' name. She was satisfied and left the studio.

The first night of the crusade during the time of testimonies, this woman, came with others who had been saved and healed. When she got to us, she said, "Do you remember me from the radio station?" With a loud voice she said, "God has healed me of leprosy. Look at my legs. The wratches are gone."

The Lord had miraculously cleansed her. The crowd went wild praising God.

There were so many miracles on that field. A little girl, about six years old, who had never walked in her life, came to the platform all dressed up. Her shoes looked new except for the toes where the patent leather was scuffed off from crawling. She said, "God has healed me. I can walk." We witnessed and filmed her slowly taking her first steps, walking for the first time in her life. Words can never describe this. One night the Lord put it on Loren's heart to preach about Jesus coming back to earth. He said, "Why Lord? I need to preach on salvation and healing. So many people are being saved and healed."

He said the Lord witnessed to his spirit, "Preach it because I am." Matthew 24:30; 1 Thessalonians 4:13.

He obeyed the Lord and began to preach on His coming again. "...I will come again, and receive you unto myself; that where I am, there ye may be also." John 14:3.

That evening was a marvel. The first thing Loren said to the crowd was, "Jesus is coming back to earth." As soon as he said that, there in the section of incurables to the left of the platform behind the crowd control rope, they stood to their feet, shouted, and praised God. The power of God shook everyone in the whole field. So many people were saved; cripples were healed; the deaf received their hearing; everything happened. A man came to the front shaking and wanted to testify. He said, "When you said 'Jesus is coming', rain fell. When the rain began to fall, my leprosy disappeared from my body."

His skin looked like a baby's, clear and soft as it could be. Later Loren said to the interpreter, "It didn't rain last night. We didn't see or feel a drop."

He said, "It did rain. It rained on the incurables section. That's when the great miracles began happening." Surely, this has to be a sign and a wonder. It sure was to us.

Another phenomenal miracle is that of a young woman about twenty years old. She testified most of her teeth were rotten and had fallen out. She said this had given her great bitterness in her heart. She said when we prayed, the bitterness left her heart and then she noticed all her teeth came back. Loren looked at her mouth, not understanding what had happened.

He said, "God healed your gums?"

She said, "No, God gave me new teeth." Her teeth were perfect. Different ones on the platform said they knew her and that it was true, most of her teeth were either broken off or completely gone. God truly is a creative God. Nothing is impossible with Him. It is still one of the most astonishing things we have ever witnessed.

LEON, NICARAGUA

We proceeded to go further south in Central America to Leon Nicaragua. It had been locked in a bloody war between the Contras and the Communist Sandinistas. At the time we went, the war had begun to wind down. There was "Yankee Go Home" graffiti written all over the bullet-riddled walls around Leon. We didn't realize how much animosity there still was toward Americans. We set up the crusade in a field facing straight across from an Old Catholic mission in the town. On the left side of the platform was a mental institution run by nuns. Every night before the crusade started, as people were gathering, we could hear hideous tormented screams coming from inside that place. It was heart wrenching.

Before the crusade began, a radio station called us to come in to do an interview. Immediately the interviewer asked where we were from. Knowing the Sandinistas were anti-American, Loren knew where the interviewer was headed and responded that he was a representative from heaven. This annoyed the man and he belabored the point, but Loren refused to fall into his trap. He finally threw down his notes, stopped the interview, got up, and left. The people came to the crusade and were responsive turning to Christ.

One night a rugged looking Sandinista came to the left side of the platform screaming at us to get out of there, but Loren ignored him and continued to preach, although our imagination worked on us.

That night, our team joined us in the town square to get a bite to eat from local vendors. A man came and sat next to us on the bench. He kept pushing and crowding Loren trying to start an incident and intimidate us. If we would have shown the slightest fear, there is no telling what would have happened.

One day during the crusade, I told Loren we needed to go to the big mental institution to pray for whoever was so tormented. I couldn't bear hearing those horrible cries every night while deliverance was being preached so close to her.

A short nun answered through a 4x4 peephole in the large wooden door. In a gruff, unfriendly voice she said, "What do you want?" We told her we wanted to pray for the woman who screamed all the time. She turned around and left us standing at the closed door while she talked to her superior. Pretty soon, she opened the door. Several nuns stood in a large hall in front of us. We explained to them we wanted to pray for the woman who was so tormented. Grudgingly they brought the poor distraught woman out, a nun on each side of her, trying to control her. When she saw us she literally broke free and ran toward us with her arms outstretched. We prayed for her and she instantly became peaceful and glowed with newfound faith. As long as we were in town, we never heard her scream again. Praise God, Jesus alone brings deliverance from demonic torment and gives peace to the troubled soul. "And his (Jesus) fame went throughout all Syria: and they brought unto him all sick people that were taken with ...diseases and torments, and those which were lunatic, And those that had palsy; and he healed them." Matthew 4:24.

God continued to move in the crusade, but so did the devil. Most of the pastors who had initially backed us now suddenly boycotted us. We couldn't figure out why there were only a few pastors with us on the platform. A religious spirit was at work. Loren discovered they secretly criticized me for wearing makeup and small earrings. Here God saved people and miracles happened, and the religious were offended and critical because my hair was too short and I wore makeup. Loren publicly castigated the pastors for their hypocritical religious spirit. I stopped wearing makeup for the remainder of the meeting to show them that I was sensitive to their concerns and only wanted to glorify God. The next day, all the pastors were back with us along with a large multitude of people. The Lord had great victory in that place.

BACK TO AFRICA

Africa was calling again. The local crusade committee in Tanzania invited us to be a part of another crusade in Tanga. During this second crusade, heaven came down. In the mornings we would go to the crusade field for a prayer meeting and minister personally to the people on the ground. Hundreds came to hear the Bible study and the people would line up for prayer. One woman who came was crippled by polio, wearing braces on her legs. All of a sudden, we heard shouting and spontaneous joyous singing. This woman had taken both braces off her legs and lifted them high above her head as she walked through the crowd singing and praising God. It was absolutely electric.

So many great miracles happened and the attendance swelled with many more Muslims coming to Jesus as their Savior. The pastor of a large Lutheran church who worked with us called Loren into his office. He told him he didn't like the preaching on the baptism of the Holy Spirit and speaking in other tongues. Loren asked him, "Have you ever seen such a big crowd?"

He said, "No."

Loren said, "Have you ever seen so many people saved and healed?"

He said, "No."

Loren challenged, "Is the Baptism of the Holy Ghost commanded in the Bible?"

"...be filled with the spirit;" Ephesians 5:18. And ,

'Wherefore, brethren, ...forbid not to speak in tongues". 1 Corinthians 14:39.

He said, "Yes."

Loren told him, "I'm not going to stop preaching and demonstrating what's in the Bible." That silenced him. "We ought to obey God rather than men," Acts5:29.

IRINGA, TANZANIA CRUSADE

From Tanga we travelled to southern Tanzania to a town called Iringa. We preached there on a crude platform made of unfinished wood, but the power of God that was demonstrated there was anything but crude. This city had the reputation for murder and suicide. The first night of the crusade, there were several thousand people present. Loren opened the Bible and calmly began to read from the book of Matthew. To our surprise, the demon-possessed began to cry out. At first he stopped and commanded in Jesus' name for the devils to come out, but there were so many possessed, he decided to keep preaching and just let our workers carry them out and handle them. We realized without a doubt it is the word of God that is so powerful. It is not our words. We have never seen so many demon-possessed delivered in our life. Loren knew when he couldn't locate me I would be in the ICU. This area had walls of burlap with no top. We should explain here that in every crusade we would set up a separate area where our deliverance teams could take the demon-possessed. Mark 16:17. I loved working to bring deliverance to people and see Satan's power broken off them and also to video their wonderful radiant faces. It was inspiring to see the before and after look on the faces of these precious people that Christ had set free. "If the son therefore shall make you free, ye shall be free indeed". John 8:36.

There was one large African mama who picked up a demon-possessed man all by herself and carried him, kicking and screaming, into the ICU pen. You have to realize it was the demons inside him that made him kick and scream and not his natural self. They wanted to stay in possession of the body. This ICU worker was determined to get this man delivered. All she knew was she hated the devil and the Lord had given us authority over them in His name and she wanted to see people set free from this demonic power.

She and another worker commanded, in Swahili, *"Toka, katika jina la Jesu."* ("Come out in the name of Jesus.")

The devils in the man screamed, "No," so she slapped him across the face and said *"toka"*, Come out". The man gave up his demons. This is exactly what happened. It's a power struggle between good and evil for the possession of a body to use. "Then went the devils out of the man...".Luke 8:33.

As Loren preached one night, the crowd on the whole right side of the field facing him took off and ran away from the crusade. We had no idea what happened and Loren just kept preaching. Soon the crowd turned around and ran back. Later we got the story. A man was demon possessed and was in the ICU. He decided he would keep his demons, so he got up and ran out, trying to escape the ICU workers. The team chased him down like at a football game, tackled him, and brought him back to the ICU, and cast the devils out of him. What a night.

Loren had warned the local witchdoctors because they had threatened us, wanting to keep control of the people. "Don't even try to put a curse on me or anyone here. Christ has redeemed us from the curse. You can't curse us or what God is doing here." "Christ hath redeemed us from the curse...". Galatians 3:13. We had so many salvations in that place.

MOSHI

We left Iringa and went to the northern part of Tanzania to visit a crusade in Moshi. Loren was not the preacher, but we were invited to sit on the platform as special guests in this meeting. After preaching, the evangelist asked him to come to the front ramp area and join him while he prayed for people to be healed.

There was a mother who brought her two small babies, under-nourished and maybe two to three years in age, demon-possessed and crippled. This by no means is to say all crippled children are demon possessed, but this situation was obvious. They were like two little spiders hanging on their mother's arms, flailing about as if they tried to climb a web. The mother held her small crippled boy out to Loren and shouted in her mother tongue, "Heal him." Of course Loren is not the healer and normally we pray a mass prayer in a large crowd.

Loren handed the boy back to the mother but she kept yelling out to him and handed the child back to him again for prayer. Loren said he didn't understand her language, but knew what she was after. This went back and forth and finally he realized she would not be denied. She put a demand on her covenant with God. 1Peter 2:24 tells us that by Jesus stripes on the cross "we are healed ".

When he saw the mother's great faith, he took the small crippled boy in his arms and commanded in Jesus' name for those crippling spirits to come out of him, then put the boy down; instantly he walked normally. She put her daughter up next and God healed her, too.

Our God loves the little children. The physician Luke tells of the situation where people were bringing their infants to Jesus and the disciples rebuked them. "But Jesus called them unto him, and said, Suffer the little children to come unto me, and forbid them not: for of such is the kingdom of God". Luke 18:16.

While in Moshi, we were invited to preach in a large Lutheran church for its One Hundred Year Anniversary. The German Lutherans had come to colonize this area a hundred years earlier. They dressed Loren in the white collar and put a purple robe on him. He was quite uncomfortable with this religious attire. This turn of events amused me and I enjoyed watching him grit his teeth as they were dressing him in this attire. He was a rugged type of man and we were now very used to "bush" and village work and the religious atmosphere and ritual were new to us. They led me to my seat and led Loren up a spiral staircase to the preaching pulpit, high above the people on one side of the church of two thousand members. Now it made no difference what he wore, he began to preach the Word of God with boldness and gave an altar call for people to be saved. Literally hundreds responded to Jesus call to be born-again. The elder of the church told us afterward that this was the first altar call ever to be given in the one hundred-year history of this church. Christ died for all but we must individually receive that gift. "Except a man be born again, he cannot see the kingdom of God." John 3:3.

I was given a special cloth commemorating this one hundred year anniversary event and we were so happy that the simple gospel was so powerful to bring so many to salvation. What a way to celebrate.

MOUNT KILIMANJARO

We traveled on to the little town of Marangu on the slopes of Mt. Kilimanjaro. It was a beautiful place with many lush green plants, coffee plantations, and waterfalls. We saw the most magnificent Poinsettia trees, ten to twenty-five feet or more high, covered with red flowers. We also saw flame trees, and another tree with purple and white blooms on them; I called them orchid trees, because the flower looked like orchids but we didn't know the names. In the states it was hard to keep a small potted poinsettia alive through the Christmas holidays, but here, well, truly, this must have been part of the Garden of Eden. It appeared everything grew ten times bigger in Africa.

Mount Kilimanjaro was incredible at 19,341 feet, the tallest mountain in Africa. Kilimanjaro sloped gently up, at least on the side we could see, but on most days, the summit was hidden in the clouds. We loved how the Africans phrased things. The local people said, "The mountain is shy today," when it was covered with clouds. We considered climbing, but soon realized we didn't have the clothing, equipment or the know-how. We met mountaineers from all over the world who came to climb the mountain and they carried packs equipped with the proper climbing boots and warm clothing. They wore shorts and t-shirts for the first days and then changed into true mountain gear and clothing to handle the cold and snow. We soon realized climbing Kilimanjaro was only for people who were in good shape physically and were quite serious and understood what they were doing.

At the lodge we sat around the fire and heard stories of how people had died on the mountain because they weren't physically fit, were improperly prepared, or had tried to take the mountain too fast. The trip up and back was five days. Three days up to allow your body to adjust to the altitude and two days down. There were camps on the way where climbers could rest, eat, and sleep. It intrigued us, but we knew we were not experienced or in shape at the time to do it. I resisted the hawkers selling the 'I climbed Mt. Kilimanjaro' t-shirts; we decided it would be disrespectful to those who really had climbed to be wearing it.

We began our crusade and the small crusade platform was set at the foot of a grassy slope where people could sit on the ground. They began to come and the Spirit of the Lord moved mightily. A man in his twenties who had been crippled by polio for many years came on crutches with a full-length brace on one leg. The Lord miraculously healed him and he came out from the crowd walking around that hillside with his crutches and leg brace held high over his head praising God. It never ceased to amaze us at the love of God.

THE PYTHON

One of our most favorite things was we loved sitting and talking about the things of God in our casual times with the local people. It was on one of those visits we heard this story.

There were two young sisters, one seven and one twelve. They were walking home from Sunday school at church and stopped to sit on a log and rest. Suddenly, a python came from behind the log and wrapped itself around the seven-year-old girl. Pythons squeeze their prey before swallowing it. The little girl began to scream and struggle for her sister to help her. The older girl had no knife or weapon of any kind, but tried beating the giant serpent with a stick, but to no avail. Finally she remembered what she had been taught in Sunday school about the power of Jesus' name. She began to rebuke the snake in the name of Jesus and to command it to turn her sister loose. Amazingly, the snake loosened its hold enough for the little girl to get away and her life was saved. They ran home rejoicing the Word of God is true for this life as well as for eternity. No wonder Jesus taught us to have the faith of a little child. Mark 10:14-15.

THE LION AND THE PREACHER

We were told another story about a native pastor who walked home from a village where he had preached the gospel. He came across two lions, a big male and a lioness. The male lion came after him. Of course it is impossible to outrun a lion and he began to cry out to the Lord to help him. The lioness turned and began to chase the big male and attack him, giving the man of God time to escape. It is almost unheard of for a lioness to attack the much larger male like she did, but the preacher lived to tell about it.

GUARD LIONS

A German missionary who worked among the Masaai had built his house on stilts in the bush. The area was full of lions and he would put troughs of water around the stilts for the lions to come and drink. He said he did this to keep away thieves at night. It worked.

ARMY ANTS

Another minister in the bush cultivated a small *shamba* (garden). This area was hit by a hoard of army ants, which devour everything in their path. They were headed straight for his farm. He read in Genesis 1:28, which says God gave man "dominion over every living thing that moveth upon the earth." Based on that scripture, he began to speak to those ants in the name of Jesus to turn their path away from his place and they did. He lost none of his crop.

DODOMA, TANZANIA CRUSADE

Over a period of time we held a total of three crusades in the city of Dodoma in the central part of Tanzania. Many thousands attended and gave their hearts to Christ; however, the Lutheran Bishop of the area openly opposed us the first time we were there. He forbade the Lutheran pastors in the area to cooperate with the meeting. We come to help the local churches, but sometimes they resisted us because not all believed in the 'born-again' message of the Bible, miracles of healing or the Baptism of the Holy Ghost. They had "a form of godliness but denied the power thereof; (Christ)". 2 Timothy 3:5.

The crusade started and the bishop left for a trip. No sooner had he left town than his wife sent word requesting we come to their home and pray for their son Stephen who was dying of HIV. We agreed to go, but after hearing how sick and emaciated he was, Loren asked me not to come in, since it was a man and not knowing how bad it was going to be. I waited outside and prayed while a small group of faith filled men went in with him.

When Loren walked into Steven's bedroom, the boy removed the sheet off himself so they could see his condition. Loren said it was the worst thing he had ever seen. He was skin and bones and it looked like a lot of his skin had been savagely torn from his flesh. To pray for him, he said he had to get completely out of his own flesh. Honestly, he didn't even lay hands on him he was in such bad shape. Loren boldly commanded the spirit of infirmity of HIV to come out of him in the name of Jesus. Immediately, Steven stood to his feet looking like someone from a horror movie, but he was full of faith. His mother was shocked he had the strength to stand. He stood with his hands lifted and he praised God, but physically one could see no change. Loren left him believing God to do the impossible.

Five days later on the closing day of the crusade, his father, the bishop, had returned from his trip and showed up at the crusade in his royal purple robe and religious trappings. He asked to speak to the multitude. Reluctantly Loren allowed him, but had no idea what he would say. Addressing the crowd, he said emotionally, "For those who don't believe in miracles, I have something to say to you. Brother Davis and his wife came to my home and prayed for my son Stephen who was nearly dead from HIV. Stephen has already gained weight; his skin has healed, and on Friday he had enough strength that he drove my car all over the town." What an incredible testimony from one who did not believe and would not cooperate in this salvation meeting.

A year later, Stephen came to another crusade we held there. During the singing he came up from the back of the platform and said, "Brother Davis. How are you?" Loren asked him who he was. He answered, "Stephen. Remember, you came to my home and prayed for me and God healed me. I'm the bishop's son." He was filled out and strong. We did not even recognize him. He was totally healed and made whole. Jesus is still healing the sick today.

GAINING INDEPENDENCE

The minister who had initially invited us to Tanzania came to us and said, "Brother Davis, you need to leave this local crusade organization. They have been restricting and controlling your ministry. He said, "They have their own agenda and are using you." We had been sensing that in our spirits and after talking to them we agreed to separate and set out doing crusades on our own.

KIGOMA-UJIJI, TANZANIA

Our first independent crusade was on the western side of Tanzania in a small town called Kigoma-Ujiji, on the shores of Lake Tanganyika. We chartered a plane from an organization that was founded to fly missionaries around the mission field, but it was extremely expensive. All their planes were donated and their staff and pilots raised their own support so it surprised us at the expense but we had no choice.

The famed missionary and explorer, Dr. David Livingstone had lived in Ujiji for a few years. We had read about him and admired him greatly. He was a Scottish pioneer medical missionary sent out under the London Missionary Society in the mid 1800's and was also famous for fighting to end the Arab slave trade. He dreamed of finding the source of the Nile and was backed by the Royal Geographical Society. Dr. Livingstone was a true man of God. He genuinely loved the people and worked so hard to tell them of God's plan of Salvation. It is said he had only one convert in Ujiji but we honor and respect his work because he never gave up. Dr. Livingston died of malaria and dysentery while trying to reach these primitive people with the Gospel. His faithful servant buried his heart in Africa and sent his body back to England where he is buried in Westminster Abby. We know he and other missionaries had planted the seeds for the great harvest of souls we would reap here.

The longer we stayed in the African bush, the more we truly came to admire the missionaries who had gone before us in this wild, harsh, and treacherous land. It was extremely physically difficult for us, so it is unimaginable how bad it was for those who preceded us a century earlier. We heard stories of how the first missionaries came to Africa in ships which took months, crossing through perilous storms. They came having packed their belongings in ready-made caskets. They never expected to see their families or go home again. The fact is many died within weeks or months because of malaria, cholera, dysentery, and savage natives. When we arrived in Kigoma-Ugigi, we were shocked to see this area was now nearly all Muslim.

When we first came to Tanzania in 1988 we were expecting to minister to native Africans of animist religions. We had a big shock that so much of Islam had taken the land and the people. Our education related to Islam and its history began to slowly emerge. Oman and other North African Islamic countries would send ships down the coast of East Africa and leave groups of young men to intermarry with at least four women each and have children for the purpose of gaining control of the land thru population. Zanzibar was one of the most interesting places to visit and to see the grandeur of what once was the territory of the Sultan of Oman.

In Kigoma-Ugigi we boldly set up a small platform in the middle of town and using the local electric power to operate our small p.a. system, we started the service. The musicians and singers played and sang; then I would sing in worship to the Lord with no problem, but as soon as Loren got up and started to preach, Muslim men dressed in full white traditional gown began marching all around the field trying to intimidate us and everyone attending, to stop the service. When that didn't work, the electrical power was manipulated, intermittently being turned on and off, making his message incoherent. It was extremely frustrating. We knew they had deliberately sabotaged the crusade. After this experience, we determined that we would trust God to bring our own generator so as not to be at the mercy of the enemies of the Gospel.

BUJUMBURA, BURUNDI

Leaving Ugigi we boarded an old ship and traveled up Lake Tanganyika to Bujumbura, Burundi. Going up the Lake turned out to be very meaningful to me. I had a dream a year earlier in which we were on some type of large ferryboat and in the dream the wind was blowing and I wore a rust colored jumper dress with a white blouse under it. At that time, I did not own a dress like this. In the dream, I saw myself on the deck of an old steamer holding my sunglasses and a large sun hat in my hands. Until we got on that ship I had completely forgotten about the dream. Now a year later, we were on our way to Burundi via the overnight ferry from Tanzania and I realized everything was very vaguely familiar. The dress I wore was the exact clothing I had seen in my dream. As I walked around the deck it stirred memories and I remembered the details of the ship. In fact, I had bought the dress at a secondhand thrift store just a few months before, never even knowing we would be taking this trip or thinking anything more of the dream. This made us wonder what plans God had for us in Africa.

In Bujumbura, we met David D'harahutsi and his wife Ruth. They had a great work in Burundi. David had lived in Uganda when the terrible dictator, Idi Amin, was in power. During Idi Amin's reign, the blood of Christians literally ran in the streets. Our friend David was caught up in the carnage. Amin's soldiers killed Christians all around him, and David's own body was thrown into a bloody pile with two hundred other butchered corpses. Miraculously, he didn't die. He stayed still and faked death until the sun went down. Only then did he dare to climb out of the heap of dead humanity and fled to neighboring Burundi.

Burundi had also had bad ethnic clashes in the past between the Hutu and Tutsi tribes. The slaughtered ran into the tens of thousands. Africa was in a constant state of turmoil. David was a tall, gentle man, with a grace only God could supply under such circumstances of life. We instantly bonded in love with him and his family.

Burundi, Zaire, and Rwanda all speak French along with their tribal tongues. The week before our crusade in Bujumbura, a horrible plague of meningitis hit the city and the government issued quarantine. This was unbelievable. Now people could not move in or out of the city. The American Embassy offered inoculations to us and other ex-patriots to try to protect us and thank God, we never got sick.

The city government allowed the crusade to be held despite this outbreak, although the plague and quarantine greatly hampered the attendance. Still, several thousand attended and many came to Christ. We adapted and learned how to operate under all circumstances.

UVIRA, CONGO

We discovered Sweden had Mission houses all over Tanzania and they were hospitable, clean, and usually had hot water and, if not, they would boil it for us for bathing. We got a room at the Swedish Mission Compound in Uvira across the border from Bujumbura. A fine Swedish couple hosted us and we had our first real Swedish meatballs and yoghurt. It was great to fellowship with westerners. They were in Uvira to oversee a tree-planting project. We discovered the Swedish government paid their missionaries and most of them were not necessarily called of God, but rather did humanitarian projects.

In these years, the Congo was known as Zaire. Uvira was later to be the site of horrible bloodletting. Our interpreter in Uvira was a precious pastor who became our friend and also had a small Bible school. Little did we know, in the near future, he and his family would be among the few who escaped the horrific killing in the Congo that took thousands of lives. For three months, he and his family walked nearly fifteen hundred kilometers across Congo, hiding in the bush and eating tree bark to escape the rebels who tried to kill them. Unknowingly, we would meet him again at a future date in Goma, Congo. However, we were in Uvira now and saw the whole town shaken by God's power.

BUKAVU, CONGO

From Uvira, we followed the mountain up to Bukavu, Zaire/ Congo. It was a beautiful though broken down little mountain city, which had an African flavor of the Belgians who had left it after colonization ceased in the early 1960's. One incident that is especially memorable is, by this time, we were hungry for familiar American foods. We stayed six months at a time and one day Loren woke up craving a banana split. I laughed in unbelief when he said he was going out to look for ice cream. He had an appointment with a pastor and left me to sort us out in this new home. When he got back a few hours later, he had a bag of African strawberries and a small tub of choco- late and strawberry ice cream. After the serious meeting with the pastor he had told him about our craving and he laughed and showed him a place where the Belgians had left an old ice cream maker. Wonder of wonders, the Africans had continued making ice cream.

Now we had the joy of knowing if God could provide ice cream in Bukavu, He would bring this town to hear the gospel of Jesus Christ. Nothing was too hard for *Him*. This crusade was a struggle, particularly because we were ill equipped and ill financed but, yet again, God showed He was strong as we were faithful to preach His Word. A touching thing happened when we arrived at the crusade grounds one night, A little woman came up to me holding an egg in both hands. She gently put it in my hands as a gift. This was so touching because food was so scarce. We had built the platform out of very expensive local lumber and had promised to give it to the local church when we left. That night, before the last amen was said, people dis- mantled the platform right out from under our feet and took the lumber. We were still on the platform with the instruments and musicians. Suddenly, one side of the platform went down with a thud and we had to scramble off for safety. People were so desperate.

MOUNTAIN GORILLAS

We went north to Goma, Congo to do another crusade in that area. The pastors in Goma encouraged us to visit the Kivu Forest while we were in the region and see the famous silverback gorillas. This was in the late '80's, still very remote, not like today which has been modernized and made into a tourist stop. Some of the Pastors agreed to accompany us. We had seen "Gorillas in the Mist", the Dian Fossey story, and were fascinated to learn we were actually in the same vicinity. We visited with a doctor in the area who told us of cases where the mountain gorillas had violently attacked people, at times severely biting them or just picking them up and throwing them. She said in this area, tending to people injured by gorillas was not uncommon. We hired a local guide and soldiers carried high-powered rifles and standard safari gear. We followed trails through the jungles, across swampy land and up the mountains tracking them by their spores.

After a good six hours of hiking, we heard a mighty bone-chilling roar. It was one of the great apes we had looked for. The soldiers had instructed us before we started that if a Gorilla roars, the next thing he might do is charge toward you, but normally they will stop just short of you. They do this to show dominance. We were instructed if this happens, we should look down and let our shoulders and arms go limp at our sides and act submissive. Most importantly, we were told not to run, because this would provoke an attack. Now, the scenario he warned us about was happening to us. We faced a great silverback male gorilla. His shoulders were unbelievably broad; his neck wide and thick; and he had fire in his eyes. We were face to face with him in the jungle with nothing between us. Our soldier guide kept his gun at ready in case of an attack.

Without warning, the silverback charged us. Unknown to us at the time, behind a little mound to our right were two baby apes playing next to their mother. We saw them later. The silverback charged us to show his supremacy and to defend his family. Although we were told to stand still and not run if something like this happens, it is easier said than done. All the natural instincts in us wanted to get out of there as fast as we could, but we conquered ourselves and stood still, eyes down and in a limp submissive posture.

The askari stepped in between the great ape and us and shook his gun toward the gorilla. Fortunately, he backed off. We backed slowly away, far enough to give the gorillas their space and watched the great giant and his family as they went about their daily lives. It was incredibly marvelous to see the creation of God. It was also sobering to think a vegetarian diet could make something that big. It was one of the greatest adventures of our lives.

GOMA, CONGO

We arrived in Goma, Congo in 1994, just two weeks after the horrific Rwandan genocide that took nearly a million lives in brutal bloodletting. Tens of thousands of Hutu refugees fled across the border of Rwanda into Goma, Congo after the Tutsis got the upper hand. The terrain of Goma is solid black molten lava. In fact, a volcano overlooking the city rumbled during our entire stay and gave off a bright orange glow at night as it threatened to blow at any time. The people told us every ten years it erupted and it was a year past due. We stayed ministering to the people.

It was difficult to get the crusade platform level because the solid lava caused uneven ground.

We wanted to minister in the refugee camps that had cropped up overnight outside of Goma town, set up by the U.N., but the Hutu in the camps sent word back that if we came, they would kill us. We did not know at the time that the refugee camps were packed with soldiers and people who were involved in the genocide along with the innocent running from the conflict. The U.N. and every major western nation sent relief flights around the clock into Goma at a hastily lengthened airstrip. The citizens of Goma quit working when the relief flights came and got in the relief lines with the refugees. It was impossible to distinguish between the refugees from Rwanda and the citizens of Goma. Everyone wanted and desperately needed help.

It was a nightmare. Not only was there the refugee problem, but also the people from the villages around Goma came in from the countryside to avoid a terrible plague of cholera that hit, along with everything else. Three thousand died of cholera every day for weeks. Every morning the corpses were stacked along the streets like cord wood and the trucks would pick them up and dispose of them. I kept a journal, my *Rwanda Diary*, and tells of our three months in this area during this terrible period.

This is the environment we were in as we prepared to minister the love of God to these hurting and dying people. The Bible says, "When the enemy comes in like a flood, the spirit of the Lord will lift up a standard against it." Isaiah 59:19. We used the same Congolese interpreter again. The meeting started off slowly. People were so traumatized it was hard for them to receive hope.

The second night, the police found two bombs planted on the crusade grounds, but praise God they were located and disposed of before the people arrived to the meeting that night. Many did come to Christ and received salvation. "That ye may know that ye have eternal life". 1John 5:13. His word gives us assurance.

We were dangerously low on finances and because of the war and the tremendous influx of people, prices had greatly inflated. We finally had to sell some of our electronic equipment to be able to get our team back down the mountain and out of the Congo. Thank God we did eventually get everyone out. "No temptation has come to you but what is common to man. But with every temptation God has made a way of escape". 1 Corinthians 10:13.

CAUGHT IN THE MIDDLE

We went back to Bujumbura, Burundi to check on our friends David and Ruth D'haduhutse. When we arrived we found there was much going on in Bujumbura between the Hutu and Tutsi, a carryover from the fighting in Rwanda. It was eerie how the battle evolved. It was quiet most of the day until about four in the afternoon when the shooting began. We stayed in a friend's apartment on a hillside overlooking the city, which in other circumstances would have been lovely. The hills were beautiful and covered in hot pink and purple Bougainvillea We would literally sit on the porch and look down and watch the war in the early evening. The shooting went on all night and then subsided in the morning. It was an unspeakable feeling watching and hearing the bursts of gunshots. We prayed and prayed for those we knew were dying in front of us. We knew people were being killed, but after a while it became surreal. At first it was hard to sleep, but after a few days, as unreal as it sounds, we adapted and slept through the night. It is hard to describe. It felt like you can only be aghast so long, and then your emotions shut down. Despite the war, through our friends we were still able to support and undergird the church and the people by meeting in the dark to pray together.

The wars made traveling difficult and perilous, everyone was having a lot of trouble with immigration at the airport but we were able to get out of Bujumbura and caught the old German ship and sailed back down Lake Tanganyika to Kigoma, Ujiji. Lake Tanganyika is said to be the second deepest lake in the world at about 4800 plus feet at the maximum depth. You could somehow feel it. It was like an enormous mountain on the underside just like the mountains visible to us above water. And, it was full of beautiful fish of every deep color God could imagine.

Sadly, several years later, we found out our good and dear friend David had died in a terrible plane crash when he and twenty five other pastors were going back into the Congo to take the Gospel. This devastated us, but David was a great man of God. The truth is, so many have paid the ultimate price to win the lost but that is what Jesus did for us. We will meet again in heaven.

Chapter 4

Impala

Returning home to the states was not an easy transition after living in the Congo and Burundi in a war zone. The reverse adjustment of coming home felt more difficult than being in the Congo in the middle of a war. It was simply two different worlds. We were shell shocked.

One of the blessings of being home was that my dad lived only about a half hour from us. He retired out in the country close to another lake and enjoyed puttering around and fishing. My mom had already gone on to be with the Lord. He was very gifted and imaginative working with his hands and had built furniture and other things as a hobby when I was young. Daddy had built my boys a go cart and in these later years he and my younger son had even built a huge metal dinosaur out on his property just for fun. Since our new home was close to him, we were able to visit him regularly when we came home from overseas.

I had prayed for my father to be saved since I was in my early twenties. My parents had taken my sister and brother and I to church but he had a lot of anger. The Lord had spoken to me to show him a lot of love, which was easier now since we lived fairly close by. One evening, he asked us if he could borrow some money to fix his pickup truck. He followed by saying he would pay it back, a little at a time from his small income. I knew this took a lot of courage for daddy because he was a proud man and would never have asked us if there had been any other way. It was hard for me not to jump right in and say yes before letting Loren answer. We were also still growing in our marriage as well as ministry life.

Now, about this time we were asking the Lord for our own plane, a DC-3, to use in our missionary work in Africa. The money we used to hire planes and vehicles to get us in and out of the bush could be put toward our own plane. Someone had already given the ministry a thousand dollars to start the fund. We hardly had any personal money and only took a bare minimum from the ministry to pay our home bills like water, electricity, and food. Loren told me later that the Holy Spirit spoke to him, "Give him the money out of the airplane account."

Mentally a battle was going on and Loren said, "Lord, that money is designated."

Then he sensed the Holy Spirit speaking to him, "That money will get awfully lonesome if I don't move for you. You need a lot more than one thousand dollars and your father-in-law needs ministering to."

Immediately he got the message. Loren turned to my father and said, "Pop, we're not going to loan you the money, we are going to give it to you."

Now, Daddy was shocked. He said, "No, I will pay you back a little at a time."

Loren said, "Pop, we're believing God for a plane to carry our team, platform and all our equipment so we can preach the Gospel all throughout Africa."

Reluctantly daddy accepted the gift, being a proud man. Our faith in the Lord's provision had increased and this gift was a massive step for us. Loren continued, "Pop, we need a much bigger miracle than you do. We're giving you the money in faith."

BIG MIRACLES

A little over a week later we spoke in a good friend of ours church in Colorado. Pastor Greg and his wife had been with our ministry from the beginning and always encouraged us in the Lord. We would be there for three days of services. One night we mentioned the need for a big plane to carry our crusade equipment, us, and our team speedily throughout Africa to take the Gospel. This would greatly accelerate the harvesting. We preached that week from Sunday to Wednesday and on Wednesday night, Loren felt impressed to have a time of answering questions about our work in Africa. There were less than seventy people that evening in this small farming community. At the end of the service, the pastor made a low-key appeal for our ministry and this vision for a plane and passed the offering plates.

After church that evening, as usual, we went to the pastor's home for a late dinner. His wife handed Loren a check from the church. He thanked them and put it in his pocket which was his normal custom, but she ordered Loren to, "Look at it." Her eyes were bright and she had a big smile on her face. He was shocked at her command but obeyed and took the check out of his pocket and looked at it. It was a check for twenty-three thousand dollars designated for the plane. When we saw the check, we all started praising the Lord and the pastor's wife and I ran around the kitchen shouting and praising God. It was an answer to prayer. It wasn't our project, it was God's. Believe me, the men were shouting too. That was more money than we had ever been given in an offering in our life of ministry up to that time. That Wednesday night at church, one couple had lovingly given a check and designated twenty thousand dollars for the airplane. The reason I was so excited was that it was nearly 23 times our gift to my father that we had given one week ago. For me it was a sign of two things; that my father's salvation was close and that God was faithful to provide for His own work of the ministry.

I had been praying for my father for many years and had asked the Lord to show me the key to his heart on how to pray. In the Congo I had a revelation about this while walking the trail to teach in a small church one day. The lava ground I was walking on was, in the Lingala language, *mabanga , hard stone*. I had to be very careful not to fall on the slippery ground. Instantly I had a revelation that my dad's heart was like the lava rock, *mabanga*. The fires within the volcano would get so hot there would be an explosion and spew *mabanga* for miles around. My dad had a tough upbringing and suffered a lot through many trials in his life and as I thought on those things, I realized how a heart could be hardened just like this ground which had a continual flow of red hot fire over the years. From that time in the Congo I began to pray my dad's heart would be softened to become a heart of flesh like the Bible said, soft, so it could receive the seed of the Word of God. "A new heart also will I give you, and a new spirit will I put within you: and I will take away the stony heart out of your flesh and I will give you an heart of flesh". Ezekiel 36:26.

As soon as we got home from Colorado, we went to see my dad. I loved to pet on him and always made a habit to stop at the Dairy Queen first and bring him a vanilla shake, which was his favorite. It was a Saturday afternoon and we enjoyed watching a ball game together on TV. It was hard not to be in a witnessing mode all the time, especially with him, but we tried. We never told him how God had blessed us so mightily since we had blessed him.

He had gone thru a lot of heartache himself and had been angry with God, not understanding His ways in general though he made sure his family all went to church. However, during a lull in the game Loren asked him, "Pop, wouldn't you like to go to heaven when you die?"

He answered quickly, "No."

The conversation was over. After that, there was no place to go. We sat quietly, not knowing what to say next. We continued watching the game, but after about five minutes with the television absorbing the silence, Dad turned to us and said, "Yes, I would like to go to heaven when I die."

We were stunned. After twenty plus years of prayer I was seeing my father make a decision for Christ. We got to lead him to the Lord right then and there. This may be the greatest miracle we have ever witnessed. It came directly as a result of prayer and love and obeying God in giving when my dad had a need. Without a doubt, this opened his heart to receive Christ. The harvest of Papa's soul was more precious than gold.

MORE MIRACLES

Pastor Greg had organized for us to go around and preach at different churches in his Colorado district to help us get the additional needed financial backing for our work in Africa. The churches we visited showed little interest. One home we stayed in showed great disrespect, treating us like we were second-class citizens for being missionaries. On top of that, one of the churches we visited gave us only ten minutes to speak. I was furious because we had shortened our ministry time in Africa to return to speak in these churches to share the missionary vision directly from the field. I felt we had left newborn babes in the Lord needing sustenance to come to these cold, overstuffed churches that did not care anything about missions or winning the lost.

After the Sunday morning service at that "10 minute" church, a man came up to us and gave Loren his card. He said, "Give me a call tomorrow." The next day Loren called him and he asked us to meet him at the church. After greeting him, he opened up the back end of his truck and said, "Could you use this?" He showed us a new red 7000-watt generator. Immediately Loren said yes, thinking about Kigoma-Ujiji and our future crusade there. When we left Kigoma-Ujiji the last time, we felt sure we had failed, but decided to return there for another crusade and this time we would bring our own generator so our meeting couldn't be sabotaged. Now the Muslims wouldn't be able to turn the power off on us. We loaded it in the trunk of our vehicle rejoicing and then that man handed us a one thousand dollar offering. Loren was always full of faith, believing the best out of everything, but I was almost sure we had missed God coming back for this trip. We were awed as we recognized God in it after all and how he had led our paths. The Bible says, "The steps of a good man are ordered of the Lord...though he fall he shall not be utterly cast down." Psalms 37:23-24. I was growing.

A few days later, back at home, we received a phone call from a local church in our area. It was from a new pastor we had not met yet. The Borden's had just moved to town to take an established church and did not know us before this phone call. The man who gave us the generator from Colorado had called this church in our home town to see if anyone knew us and to find out if we were a legitimate ministry. There happened to be another local pastor, Bill Bloodworth, visiting in the church office at the time he called, so while on the phone with the man from Colorado, Pastor Borden asked Pastor Bloodworth if he knew us. He told him, "Yes, they have a great ministry in Africa." The pastor relayed the message to the caller and because of that testimony; the man sent us another check, this time for ten thousand dollars toward the plane. It's impossible for it to be a coincidence that the man called when our friend was in the office and able to validate us. It was the hand of God.

PARALYZED

While in the states, we were to preach in a small town in Texas. I drove and gave Loren time to relax and read a newspaper so it was a quiet drive. I noticed him fidgeting and rubbing his eyes but he didn't say anything to me and I knew he was very tired from all our traveling to raise support. We met the pastors for supper before the service but while seated at the restaurant across from them, the pastor's wife looked at Loren and said, "What's wrong with you Loren?" He was never a complainer and I didn't understand what she was talking about. He said he knew his eyes were blurry and he had a hard time opening his lips, but neither he nor I realized what happened to the right side of his face. I hadn't noticed because I was on his left all day during the drive and even now at the restaurant; the side I could see appeared normal. The pastor's wife saw his right eye wouldn't close and that he sounded funny.

When I looked at him full on and saw his condition, I was concerned. We prayed over him and he decided to preach that evening anyway, but struggled and slurred his words badly. We planned on returning to Africa in two weeks. Now the right side of his face seemed paralyzed. We had always believed the promises of the Bible for our health and rarely let anything slow us down or stop us. I had gone thru a fearful time with my own health because my mother passed away from a heart condition so I was tormented for a time that I was subject to that as well. I made myself bear down and claim the scriptures regarding healing, realizing the enemy of my soul was using that fear to keep me worried every time I had a pain. It took a long time for the word of God to sink into my spirit and drive that fear out, but with the Lord's help, I was able to get thru it. God's word says "Faith cometh by hearing, and hearing by the word of God". Romans 10:17. We read the word of God and hear the word and it builds our trust in God's promises.

We continued working now, as we always did, praying and trusting God for what was happening to Loren. We began our marriage on a thin budget and learned back then we could believe God to take care of us because we couldn't afford medical insurance and had no choice. He had been faithful.

Two days before we were to leave for Africa, Loren flew to Arkansas to preach for another pastor friend. That Saturday evening, as they went out to dinner, sitting across from him the pastor said, "Loren, you have always been so strong. What happened to you?"

He told him, "This is temporary; I will be fine. I'll preach for you tomorrow and on Monday I'm flying home and connecting on a flight to Africa. God is going to heal me. Don't worry," This was not a contrived faith, Loren was a real believer that God would do what he said he would. We are not saying not to go to doctors; we believe they are there to help. It was just that it never occurred to us to turn to them for help.

The next day he preached at this large church. Although his words slurred badly and his right eye gave only a blank stare, he told the people, "Tomorrow we are leaving for Africa, and the God who is going to save and heal the people in Africa is going to heal me." We had no doubt.

I had stayed home packing and getting things ready for us to leave the country again. On Monday morning, Loren flew back to rendezvous with me at the airport to catch our British Air flight to Africa. Loren's mother and son were there to see us off. They knew there was no way to keep us from going, so even though they were concerned, they hugged our necks and bid us goodbye as we boarded the plane. During the long flight, we confessed the healing promises of God, "By His stripes I am healed" Isaiah 53:5 and 2 Peter 2:24. On our way across the north Atlantic toward our first stop in London, every once in a while Loren got up to go to the lavatory and look in the mirror to see if he was visibly healed yet, but there was no change. This did not deter our faith in God at all. We arrived in London and nothing had appeared to happen yet. I knew Loren well and we were in agreement in our faith in God's Word.

Our next leg was toward our final destination, across North Africa heading for Dar es Salaam, Tanzania. He continued getting up periodically and going to the lavatory to look in the mirror and check to visibly see his face. He tried closing his eye and tried to move the side of his lips that were paralyzed. There was no change, but we were undaunted. We must have been about two hours to arrival in Dar when he went to check on his face again. This time his eyebrow moved a little. He came back to the seat and excitedly showed me. He said, "It's happening—God is healing me." Soon, he went to look again. This time he was able to move the paralyzed side of his lips because he could see his moustache moving. He was so excited. When our feet stepped on the tarmac of the airport in Dar es Salaam, the paralysis was completely gone and his face and eye were normal.

SECOND CRUSADE AT KIGOMA-UJIJI

We again began to penetrate deep into the jungles of Tanzania to many remote small towns taking the Gospel. This time we had our own generator for the crusades and our enemies could not tamper with the electricity. We toughened up and were more filled with resolve. We returned to Kigoma-Ujiji on the shores of Lake Tanganyika for a second crusade. The power of God was something to behold. Many mighty miracles happened. Loren was not speaking his own words but preaching the words of God. One crippled woman who had crawled to the crusade was instantly healed and walked home from the meeting. There were many demon-possessed delivered. Each night after the crusade, the ICU workers continued delivering people and screams could be heard from them late into the night as they battled for their freedom. It was a marvel sitting on the platform and seeing people pour in from every direction coming from the bush and the streets. There was a mighty breakthrough for Christ in this Islamic region. The seeds Dr. Livingstone had so faithfully planted came up with a great harvest. 1 Corinthians 3:7-8 says, "So then neither is he that planteth anything, neither he that watereth; but God that giveth the increase. Now he that planteth and he that watereth are one: and every man shall receive his own reward according to his own labour."

KIBONDO, TANZANIA CRUSADE

We continued our crusade safari to the village of Kibondo deep in the interior. By this time a band of six musicians and singers from Dar traveled with us and lead the people in singing choruses and praising the Lord. God helped us to increase not only in the size of the crowds we preached to, but also in the number of our team. It is a big responsibility caring for people. I now became more than a wife, teacher, singer, evangelist, counselor, and camera person; but now functioned in the capacity of chief accountant and "mum" to the team. Everything was a step of faith for all of us. We lived by faith, totally dependent on the Lord for our sustenance. "The just shall live by faith: but if any man draw back, my soul shall have no pleasure in him". Hebrews 10:38.

Our presence in these smaller towns was a novelty because they rarely, if ever, saw *wazungu*, "white ones", and also a ministry like ours which brought the powerful Book of Acts ministry to the bush. We had a great crusade in Kibondo. We loved working in remote areas even though the living conditions were rough.

It was here in Kibondo where a man had recently been killed by a lion in the middle of this village. Honestly, while preaching, we looked around to see if a lion might be coming. Having no weapon, the only thing we could do was use the name of Jesus. The whole village came. We still have mighty spiritual weapons through God that are stronger than any natural weapon. The Bible says, "At the name of Jesus, every knee shall bow", Romans 14:11, and lions also have knees. Our own faith got stronger by the day.

CHIEF SENGE

The crowds and miracles we had in Singida were awesome. Everything imaginable happened.

Singida is where we first met a Muslim chieftain, Chief Senge. When Tanganyika got her independence from the colonists in the early 1960's, the new Tanzanian government gave the former Chiefs government positions. Chief Senge was now one of sixteen chiefs in Tanzania. He was a trim stately man of some years and totally bald. One of his friends on the police force was deaf, and had come to the crusade, instantly receiving his hearing. When Chief Senge saw this miracle, he came to the platform with his friend and greeted us. Right there, he gave his life to Jesus.

We became good friends and he took us to his house. We walked through the dusty town to his home. He lived in an old, run down 1800"s colonial house. It was only a shell of its former glory. The porch, the house, the attic, and the trees around the house swarmed with killer bees and their hives. It took a lot of courage to even enter inside the compound because there were so many of them. Chief Senge believed bees living in his home were a blessing. If that's true, then his home was super-abundantly blessed. We bravely entered into his house and he introduced us to some of his wives. He practiced polygamy. We asked how many children he had and he told us that he had thirty sons. He didn't know how many daughters, they only counted sons.

THE BATTLE WITH LIONS

Chief Senge had been in several scrapes with lions. On one occasion, he and two other men came across three lions, two lionesses and a big male. The lions turned and attacked them. Senge and his friends had guns and eventually were able to kill the two lionesses, but not before his friends were wounded. When his friends ran out of bullets in their rifles, they climbed trees trying to save their lives. Chief Senge was left alone on the ground to fight the big angry male. Lions don't run *from* a fight, but run *to* it. Now he was down to his revolver pistol and had only one bullet left. There was no way he could outrun the lion who was determined to kill him, so he had no choice but to face him at close range and fight with what he had. As the savage beast bound toward him, he had to control his emotions and wait with nerves of steel until the lion was right on him. The charging five hundred pound lion came to finish him, but at the last second Chief Senge shot him between the eyes. The beast dropped dead instantly at his feet.

THE MAN EATER

As mentioned earlier, after colonization, chiefs were made government officials and Chief Senge became the game warden of this province. On one occasion a man-eater had come into his area and killed and ate several people. It was Chief Senge's job to kill him. One day, he was told a lion had killed another man. Knowing lions sometimes return to eat their prey at night, he went out to the man's carcass. He removed the man's partially devoured body. With his rifle in hand he laid down where the slain man was found and patiently waited for the lion. Around four a.m., the lion returned. Senge rose up and killed him at point blank range. Chief Senge was one of the toughest and bravest men we have ever met.

OUR IMPALA HUNT

On our visits, Loren told Chief Senge he loved to hunt so he took us out hunting with several of his friends. He let Loren use his gun, a .243 with a small .22 scope. Believe me; we weren't going for lions with this. We hunted for Impala, a beautiful reddish deer-like animal with two twisted horns. They are a favorite of cheetahs. We rode with the Chief and his team in the back of an old grey Land Rover pickup. I was game for almost anything, which made it a lot of fun for both of us and strengthened our marriage. We lived twenty four, seven doing everything together and this was a part of our life in the bush. I had to make a decision to commit to it or go home and I knew Loren wouldn't do well without me. I learned to like it as part of our ministry to these bush people.

We stalked the herd in the bush, but when they saw and heard our pick up, they took off running with their high majestic leaps. Loren had a hard time getting a shot because of the dense woods, but also the truck raced with the herd and hit every obstacle in the bush.

Finally, a beautiful buck stopped at quite some distance and faced us. It would have been a difficult shot since Loren could not see his body, but he decided to go for it. It was a high-pressure situation because Chief Senge and his friends who were all great hunters, along with me, his wife, were watching. Loren has never been faint hearted. Looking through that .22 scope didn't give him a smart target and he had never shot the Chief's gun before. He leaned across the top of the cab and sighted him in. He squeezed a round off and immediately the impala's legs flew out to the side and he fell hard to the ground dead. Chief Senge and his friends along with me began shouting and celebrating. Chief Senge said Loren was like Mike Tyson, a one-punch knockout. I shook my head at Loren and said, "That was the Holy Ghost." For sure we knew the Lord helped him, but he told me, laughing, "at least I pointed the gun in the right direction". It was great to share this with him and I told him later I was right behind him and prayed for him to make that shot since Senge was the Chief and it was a matter of honor. That difficult shot caused Chief Senge to respect Loren as a man and cemented our relationship. We roasted and ate the impala over an open fire pit and celebrated the day's hunt. Villagers came and we shared the kill with them. Chief Senge liked me and told Loren that I was the only woman he ever knew who would go on one of these hunts in the bush with the men. I think it must have been a cultural statement because wives have a different position in his world but I appreciated his thoughts. He said he liked me because I seemed not to be afraid of anything, not even camping out in the jungles. That was God working in me.

We stayed in Singida for some time and it gave us an opportunity to have many more outings with the Chief. My favorite memory of him was when we came across some fresh elephant tracks. He stopped and picked up a handful of elephant dung. I cringed, a natural reflex for an American woman who carried hand sanitizer with me everywhere. Crumbling it in his bare hand, he said nonchalantly, "Elephant, three days" and pointed in the direction they were moving. My eyes got big, then I smiled and we all kept walking as if nothing unusual had happened. We learned a lot about tracking animals in Africa from this great hunter and loved every minute of it. God had given me a special grace to live this life for the sake of reaching people with the Gospel, but I quickly learned to love and appreciate this wonderful gift of being placed "here for such a time as this". Esther 4:14.

THE CAPE BUFFALO HUNT

We camped in tents deep in the bush outside Singida town. On this trip Loren had brought his improved 30'06 rifle, which made it similar to a .300 Weatherby. It was quite powerful and had big knock down power. At the time we didn't realize it was no match for a lion or Cape buffalo. They don't die fast. One day while staying in tents, Chief Senge took us for a walk in the bush. Loren didn't take his gun, because Chief Senge was there and the chief wasn't afraid, so we assumed it was all right. We came across a recent kill by a lion and stopped assuming no matter whom was with us. We needed the Lord every step we took. Our trust cannot be in man or weapons alone.

That evening he and Loren were determined to go hunting for Cape buffalo which are some of the most savage and dangerous beasts in Africa. They have been known to attack and can be very aggressive when confronted. We went in the chief's old beat-up pickup and backed into some bushes about seventy-five yards from a waterhole. I don't like to miss the action though I was nervous. We arrived there about four a.m., waiting for the buffalo to come to drink. Our pickup was hemmed in with brush all around us. Suddenly it was like Loren came to his right mind. It dawned on him that his gun was too light to hunt a Cape buffalo. He began to remember hearing stories that hunters normally use at least a .404 or .500 caliber big game gun to hunt buffalo.

It was dark with little moonlight, and Loren knew if his shot wasn't perfect and the buffalo was wounded, he would come after us with a savage fury and a vengeance that could kill us all. Thankfully Loren told Chief Senge, "I'm calling the hunt off. Let's get out of here now." I thank the Lord as Loren was getting older, wisdom was kicking in. When he was a little younger, he might have tried to kill that buffalo with a spear. I'm truly amazed we have survived ourselves had it not been for the Lord. We did spend many nights in tents in the middle of lion country, however. We may have been called a lot of things, but no one has ever accused us of being cowards. We didn't hold back for the Gospel's sake.

KAHAMA

We continued working in the remote towns. Despite the usual living conditions, we had a wonderful move of God in Kahama Tanzania. During this meeting we met a special young man. He was vibrant and truly loved the Lord and the Gospel. He was moved by the power that the Lord used thru Loren in the crusade and by my daily Bible teachings on the grounds. Every morning I would teach for hours and he became like a little bird with his mouth open, drinking in every word. He was a technical schoolteacher, but wanted to go into the ministry. Loren and I took an interest in him and mentored him. He received the baptism in the Holy Spirit in the morning teaching sessions and developed into a fine man of God. Years later we came back and built a church for his village.

NZEGA, TANZANIA CRUSADE

We were so poor the first ten years of our ministry we could only stay in the native small town butchery/hotel if we weren't camping in the bush. That's literally what they call them. The butchery and the hotel were in the same building. For you to

86

grasp what we are saying, "hotel" can't be defined the same as westerners know it. Most of the time if you saw "hotel" it would only mean a small place to get snacks and hot tea. This one was connected to the butchery and was wall-to-wall concrete floors, old paint, and dirty walls. The bed for both of us was usually a single or twin size, pushed together with a few slats sporadically put across it and no plywood covering it. The mattress was a piece of foam two inches thick. Believe me, there were some long nights. You wished for the day. The floors were made of thin cracking concrete that had dirt holes in them but we thanked the Lord for protecting us with a roof and walls.

Every morning our alarm clock was "chicken-killing time." The butcher lined up the chickens and killed them right at sun-up outside our window. We had never heard such blood curdling screams. The chickens knew they would get it. As soon as they picked up the next chicken he screamed and knew he was about to be killed having seen his comrades go down. This is not the most pleasant way to wake up. We also knew at lunch or dinner we would have to face these chickens again. It didn't make eating something we looked forward to. In the states, eating is an event, but here it was something to put in your belly to strengthen you for the day. Most of these chickens died in vain as they were skinny and tough. Without strong teeth, we would have starved. Most natives only have one meal a day of maize porridge known as *ugali*. In many places, we were fortunate to be able to have these chickens.

TABORA

We stayed in an old colonial hotel in Tabora built in the early 1900's during the hay day of the European occupation and it appeared it had never been painted since. Our room did have a bathtub and a sit down toilet, which was a major luxury even though it didn't have a seat. One thing we learned was that even though modern conveniences appeared to be there, it didn't necessarily mean they worked. We turned on the tap in the tub preparing to bathe since we had been on the road for six weeks and hadn't seen a bathtub or shower for a long time. In the bush and in small towns we would bathe by heating water over some coals and basically taking bucket sponge baths.

As we turned the faucet on, dark brown, muddy water filled the tub. After a few minutes of running the water, it became evident it wouldn't get better. We finally gave up hope in great disappointment. We couldn't see the bottom of the tub. Neither of us were about to get into that. Instead we took a sponge bath with that muddy water. In reality, we exchanged old dirt for new dirt, but at least it felt like we did something and at least it was wet. An unexpected hard rain came that night and I jumped up out of bed and immediately handed Loren the buckets to take outside to catch the rainwater. Now, we had clean water to bathe in. We felt like we were in the lap of luxury, God's shower.

KILLER BEES

We boarded a train from Tabora headed back for Dodoma. Our team had gone ahead to prepare the meetings. It was a long filthy hot and rough ride. We couldn't open the windows wide enough for ventilation; and sleeping was impossible. After many torturous hours we finally arrived in Dodoma to visit some missionary friends from Canada and to hold another crusade. We had met them at our crusade a year earlier on our first trip to Dodoma. They had been living there for nearly ten years and had a house. When they opened the door and saw us, their mouths dropped and their first words were, "Come in. Take a bath. In fact, take two." We looked pretty rough.

One night during this crusade while a group led the singing on the platform, a large dark cloud of killer bees visited us. They swarmed the platform. In this area, normally when killer bees come around, everybody runs away out of fear. The bees are aggressive and deadly. I saw the swarm coming as I was filming the singers and began to rebuke them in the name of Jesus. I didn't want the crowd to panic and we all joined in faith on the platform to divert them spiritually. Our singers never stopped singing and not one person ran away. The big swarm of bees dove at us as if they would sting us, but then at the last second swerved away like there was an invisible shield around us. Miraculously, not one person was stung. It was a sign and a wonder. Later, witchdoctors confessed they had sent them to stop the crusade, but when it didn't work they gave up the fight. It was another opportunity for our God to show Himself strong. "For the eyes of the Lord run to and fro throughout the whole earth, to shew himself strong in the behalf of them whose heart is perfect toward him". 2 Chronicles 16:9. We are not perfect but He is.

The meeting was wonderful and everything exalted Christ and told of His saving and miracle-working power. Many people accepted The Lord as their saviour. One night we noticed a little crippled girl behind the crowd control rope. After prayer, she tried to walk, but to no avail. Night after night she came and tried to walk after prayer, but nothing happened. Then, on the last day of the crusade, we noticed her again. This time she was dressed up in a white dress and white hat with white shoes. She stood feebly and hung onto her mother's dress. We could tell she had come to get her miracle that night. After hearing the Word of God and the prayer, she turned loose of her mother's skirt and tried to walk again. She was wobbly at first, but you could see the faith in her eyes and her determined expression. Before long, she was walking well. Jesus had healed her. The crowd exploded they were so happy at the demonstration of God's mercy and love.

CHICANERY

As we explained earlier, we believed the Lord for our own airplane to be able to get to tough places to preach. We had invested in every lead possible as well as letting the plane fund build up. It was a slow process, but we had a lot of partners who could see the vision with us and they tried to help us as best they could. All in all, we knew it would be the hand of God to be able to get us what we needed. We had decided that a DC-3 would be the best plane for us, because it would be large enough to carry our equipment and team and was well adept at operating on short dirt strips we would use.

Loren stopped in and visited the manager of a particular Christian aviation organization in Dar about registering our plane in Tanzania. He said this would be difficult for us to do on our own, but he suggested we put the plane in the name of their missionary aviation organization and that would make it simpler. We did not want to lose our freedom by coming up under another ministry so we declined the suggestion.

Our advance man from Tanzania suggested we register our ministry in Tanzania. This would help solve our problem. We tentatively agreed and sent him a copy of our constitution and by-laws. We corresponded with him from the states but when we returned to Tanzania, we found he had not followed our instructions and showed us a totally different constitution and by-laws, which gave us absolutely no protection or authority over our own ministry or property. Furthermore, he had a list of thirteen men on the board, most of whom we didn't know. Realizing if we brought a DC-3 and other equipment into Tanzania, they could take a vote and take everything, we rejected this plan, and began to continue to look for another way.

One thing that bothered us and added to our deepening concerns was we had also sent funds to him to print posters for our meetings. We did not get the amount we paid for and what we got was inferior in quality. We began to wonder what happened to the money, but still we gave the benefit of the doubt. Because of the cultural and language differences we erred on the side of misunderstandings. This was to be part of our African education.

MONYONI

This particular trip was treacherous in another way. We found wherever we went, a Finnish evangelist based in Tanzania, would either go before or after us and try to sabotage our crusades. This was done on three successive occasions. Our meetings had been set up with the pastors long before this man had entered the picture. He sent his advance people into town right in front of us to persuade them to work with him instead. We did not understand what was going on and had no idea why, as big as Tanzania was, this man would try to hinder us. It was like he was a territorial lion. There were so many towns that needed the Gospel. We even had to cancel one of our meetings because the pastors said they could not work on his and our crusade at the same time on the same date. They said they didn't want to lose either one of us helping them, so they were indecisive. This evangelist's advance man would try to come in on top of us literally everywhere we went. They wouldn't meet with us either to discuss any problem they felt could be resolved.

This first happened in Kigoma, and then again in Monyoni. Although we cancelled Kigoma because of this, we decided to stand our ground in Monyoni. It appeared this evangelist was given the assignment to stop us in Tanzania. We set up our meeting right across from a Muslim mosque. God moved this small town and the response to come to Christ was tremendous. From Monyoni, we returned to Singida for another crusade. That same evangelist tried to sabotage us there as well. In fact, he was successful. We were mystified at this behavior. This was our first major introduction to the "Spirit Wars" that go on *inside* Christianity.

SHOWDOWN IN THE BUSH

Loren still had his .30'06 rifle on this trip to do some hunting. The airlines had secure ways of transporting weapons as long as the country you were going to had hunting privileges. Of course we did all the correct paperwork and things traveled smoothly. Tanzania allowed game hunting as I told you about Chief Senge. We were now in another region and one of our friends, Samu, worked for Tanzania's Wildlife Service and offered to be Loren's guide in this area. He took care of all the permit issues and we started out. We borrowed an old Land Rover pickup and headed for the bush, but the first village we came to, we met up with Ndovu, the eldest son of Chief Senge. Ndovu asked us for a lift as his pickup had a breakdown. Before we knew it, he took over driving our Land Rover. He had his father's .404 big game rifle with him and soon we found ourselves in a strange village. We had no idea where we were. Ndovu said he needed to borrow the Land Rover for a few minutes and he would be right back. He was presuming on our friendship with his father and it was awkward but we decided to trust him. He dropped us off under the only large shade tree around. Fortunately we had some portable cots and chairs with us and I had arranged for a couple of women to come with us who were to be our cooks because we had planned to stay in the bush awhile evangelizing.

We waited for Ndovu to return, but the waiting was in vain. Hours went by and our tensions grew. We tried to rest on the cots, but would have to move every few minutes to keep the sun off of us because the tree didn't have many leaves and provided little shade. The natives there were curious but aloof and watched us from a distance. There was a huge hole in the side of the tree and killer bees came in and out of it and that bothered us. Finally a man came over and told us there was a large, aggressive snake that lived in the tree. He said that was the reason everyone kept their distance and advised us to be careful. That compounded our problems. The afternoon wore on, and still there was no Ndovu.

Late in the afternoon a big group of men came toward us menacingly. Samu looked the situation over and decided it didn't look good. He spoke to them and let them know we meant no harm and that our vehicle should be coming back at any time. We knew nothing about this village or the people who lived there. Loren took the .30'06 rifle out of its case and laid it across the cot to show them we were armed. When the villagers saw this the men halted their advance on us. Loren felt he had the responsibility of protecting us and our small team.

By now we realized Ndovu may have hijacked the Land Rover. We were all annoyed and didn't know where we were or what we would do that night since our tents were in the possibly stolen vehicle. Finally, after waiting eight long hours, Ndovu sped back into the village in a cloud of dust. The back of the Land Rover was full of fresh-killed zebra meat. Now it was obvious he had used our vehicle for hunting and sold the meat to the various villages that afternoon. Loren ran over to him and confronted this huge, muscular 6'6" mountain of a man. He said "What do you think you are doing taking our vehicle like that?" Ndovu was a giant and a professional hunter. Now, the people of the village started running over to the vehicle where Loren stood up to Ndovu. When I saw them coming, I ran toward the villagers with my arms out stretched and said, "Stop in the name of Jesus." I didn't want anyone else getting involved in the volatile situation. They stopped in their tracks, unable to believe I was so bold.

At this point we saw Ndovu taking the battery from our vehicle and putting it in his old vehicle that he had apparently abandoned earlier at this village. Loren went over to his vehicle, took our battery out and put it back in our Land Rover and walked back to the cot, where he sat with his hand on the rifle. Ndovu came over and put his hand on the barrel of the gun. They each had a grip on it. It was an extremely dangerous moment. God somehow cooled this thing off and he backed down and left our battery. We packed up our gear, got in our vehicle, and headed out of that village. Although we were deep in the bush, we were all relieved when we set up camp that night although, praying Ndovu wouldn't try to track us down for further confrontation. We never saw him again.

Chapter 5
Fish to Fur

Our ministry was always mainly outdoors and this presented a problem during the rainy seasons in Africa. We felt called because of this, to rotate our year with ministry in the Asian countries wherever the Lord opened the door.

I researched how we could physically get to Asia and heard about becoming a courier. It was 1994. I found out that we could carry documents and have our flights paid for by the companies entrusting us with important documents that needed hand delivery. We signed up and were accepted into the program.

We planned to minister in the Philippines and India and made a stopover in Hong Kong. We were invited to preach in a local church in Hong Kong by a minister friend from California who had an outreach ministry there and that was our first stop as Couriers.

Many people were already trying to get out of Hong Kong which was a part of the British colonization for more than 150+ years. It had emerged as a wealthy east-west trade center. It was due to be handed back by treaty to Red China in 1997. People did not trust what would happen to the booming economy when it reverted back into Red China's hands. They were running to Canada and the United States as fast as they could get out.

All these countries were on our heart. Everywhere along the roads in Hong Kong we saw processions for the dead with life-sized cardboard cars, houses, card games on tables with chairs including play money, along with other things. We asked what this was all about and were told this play money was "hell money" so the person who has died will have money to spend in hell. The playhouses and cars were so they would have a place to live and something to drive in hell. We were astonished at their acceptance that hell is where they went when they died and they would have a good time there. This was our first encounter with the concept of a Buddhist friendly "hell."

The Bible says this about hell: "depart from me, ye cursed, into everlasting fire, prepared for the devil and his angels." It was not prepared for people but if we, as humans, do not choose to follow God's way to bring us to salvation, then we will partake of that place also. Matthew 25:41. God does not send anyone to hell, we send ourselves by not accepting Jesus Christ as "the way, the truth, and the life: no man cometh unto the Father, but by me". John 14:6. That is what evangelism is all about, showing people *the way*.

There were Shrines and Temples everywhere. The idolatry was unbelievable. We listened carefully as our host explained the things we were observing. A prostitute stood in front of the altar silently mouthing words to a man-made god she hoped would hear and make a difference in her life. It was heartbreaking what we saw and felt our calling getting stronger to preach the saving Gospel of Jesus Christ even more than we ever had before. One thing that also struck us was that Hong Kong never sleeps. At night the streets were lit up like broad daylight and every kind of business went on as if it were daytime. The streets were so crowded and the opulence was overwhelming. Diamond and Jewelry shops, every Haute Couture name brand had a big store; it was screaming money, money, and money. The cars were all high end vehicles, BMW and Mercedes.

However, God had His people working even in this society. In one sidewalk booth, we found a huge pile of fat China baby dolls with passports attached to their arms. In looking at several dolls, I noticed that not all the passports were the same. Most looked like normal passports with the normal information you might find, but scattered in the mix were passports that said, "destination: Heaven" and had the scriptures of how to get there. What a blessing it was to realize that those dolls had been seeded in to be a witness to someone on how to get to heaven. "That if thou shalt confess with thy mouth the Lord Jesus, and shalt believe in thine heart that God hath raised him from the dead, thou shalt be saved." Romans 10:9-10. We had little money for extras and each doll was the equivalent of twenty-five dollars, but we sacrificed and bought two dolls for our only two baby granddaughters at the time. It was such a joy to realize there was evangelism going on within Hong Kong by his secret people.

We were invited to preach in the state controlled church in mainland China, but declined. The true church in China is underground. It is an atheistic nation and true Christianity is its enemy. However, the light of the Gospel shines the brightest in the greatest darkness. Men will never stamp it out. Pray for the persecuted church in China and around the world.

MANILA

We learned there was a DC-3 for sale in Manila, Philippines. It so happened we had planned a crusade there. An airplane mechanic friend of ours from Los Angeles went with us to look at the plane and check it out. It actually looked good, but when Alan took the cowlings off, he discovered the plane had no engines, only dummy partial engines. We were learning a lot about what we might be facing financially in new parts for any older plane. We were grateful for the learning experience.

A local pastor friend of ours took us to the beach area and bought us some of the favorite food of the Filipinos: fresh squid.

We watched the guys ink the squid in pails of salt water, throw them in hot water for a brief time, then take them out and hand them to us to eat. The squid still had its suction cups on its tentacles. We tried one, but honestly both of us felt like we were eating a rubber inner tube. We also discovered they ate dogs. We couldn't stand to see them on the porches after that. Discovering these culture clashes helped us understand the people better and the Lord was working on us about how we thought about things that were a challenge to our western sensibilities.

We set up the equipment for a small crusade in the middle of town. There was a lot of street activity with people milling around. The attendance was poor, but a transvestite attended the meeting the first night. He came to the platform dressed in a wig, lipstick and makeup, short tight skirt and looked like a woman. He confessed he wanted to be saved but didn't know how, so we prayed with him. Right there in front of everyone one of the most incredible transformations of a human being took place that we had ever seen. It was almost shocking seeing him after that perverse spirit left him. "…male and female created he them". Genesis 1:27. His countenance became completely masculine. What a miracle. The power of Jesus Christ changes a person inside and out.

MINDANAO

From Manila, the capital, we flew over to General Santos City on Mindanao Island. There are about seven thousand small islands that make up the Philippines in South-East Asia. Mindanao has Islamic influence and we were here to preach another crusade. Everywhere we went there were signs out in front of the businesses that said, "Check your guns here." It was like the old American Wild West. Our host took us up to a village in the mountains to look around and the villagers graciously fixed dinner for us. They brought out live fish about the size of large blue gills, slit their bellies in front of us, and put them on the grill still flopping. After a few seconds, dinner was served. We said grace over them and we were thankful.

Our crusade in General Santos City, Mindanao was small, but memorable. We had a dinner for pastors to introduce our ministry and many came out to meet with us.

The day before the crusade, the Muslims attacked another town not far from us called Ipil, in which they killed over two hundred people and did a lot of damage to the town. This put General Santos City under red alert. The authorities allowed us to have the crusade using bodyguards, but the attendance was light. One night our car was delayed getting to the crusade and everyone waiting for us panicked. They thought we had been kidnapped because apparently this was a common occurrence on this island. We didn't know that danger existed until then.

A WHALE STORY

After the meeting our young pastor friend wanted to take Loren fishing. He told us a story about how his uncle would take his small dhow out in the ocean to fish at night. He would use lanterns to attract the smaller fish, which in turn would attract bigger fish. Sure enough, he did finally attract a big fish, but it was more than he could say grace over. The water began to swirl and a whale rose up out of the water slowly and leaned over, staring at him with one big eye. His uncle put his head between his shaking legs, believing he was about to be supper. After a few moments, the whale slowly slid back into the water and out of sight. His uncle immediately returned to shore and didn't go to the sea again for months.

This time the fishing yacht turned out to be a Filipino dhow with two outriggers and a homemade sail. The dhow looked like it could have been a hundred years old and probably was. I didn't go with Loren this time. I remembered our last fishing trip out of Tanga in Africa. They headed for sea under sail on the dhow. Loren loved the sea and thought sailing was great no matter what kind of boat. He really enjoy the water and fishing of any kind. The dhow was about six to eight feet wide with outriggers attached to it on each side. The crew consisted of several Filipino men.

After a while, he told me they rendezvoused with another dhow at sea selling anchovies and they bought some to use for bait. Getting back under way, the Filipino crew began boiling water and throwing some of the anchovies in it. After a few minutes, they pulled the anchovies out and ate them whole with raw cucumbers. It was their galley and that was what was on the menu. They offered Loren some, but he graciously declined, saying "Not today. Thank you." He had brought some donuts with him from town.

Soon, he discovered, like Tanga, they were to use drop lines to fish. Fortunately, this time he had also brought some gloves with him anticipating this scenario. They got into a school of nice-sized Dorado and began to hook up. He told me what great fun it was. If he held the line too tight, the Dorado would nearly pull him over the side. Then all of a sudden it turned and swam toward the boat, which nearly made him fall off the other side because it was so narrow and he pulled so hard. He had a ball. I thoroughly enjoyed hearing about his day and was glad for him.

It was good to unwind after the stress of everything going on around us. It was 1994-95. World events were heating up.

OUR FIRST INDIA CRUSADE

The last leg of our Asia trip took us to South India. We had been invited to preach a crusade in Alleppey, Kerela State not far from the city of Cochin. There were Hindu temples and shrines practically every few feet. Loud music and bands played in the streets and Hindu festivals happened nearly nonstop. The foot traffic was thick and people stopped at the shrines and offered coconuts and other fruits to countless gods.

South India was beautiful; rice fields and Elephants and we loved Elephants. These Indian elephants were the largest we had seen in terms of actual height and it was amazing to watch them in the middle of traffic. The Indian elephant has smaller ears than its African cousin.

We had the experience of actually riding one and it was incredible these gentle, trained giants could be so seemingly docile.

In the midst of this meeting we saw hundreds accept Christ. Preaching from 1 Corinthians 15, Loren told the story of Jesus and how He was God come from heaven to be our perfect sacrifice: He came in a man's body but his blood was God's pure blood, uncontaminated by sin; the sacrifice the Old Testament spoke of and pictured by the sacrificial lamb; to die on the cross for our sins. He rose again on the third day so we also could rise and live again with Him in heaven someday. One man lifted his hands to accept Jesus as his Savior, the tears flowing down his cheeks and without anyone instructing him to do it; he took his forearm and rubbed the Hindu bindi (red dot) off his forehead. He was so grateful Jesus died for him to give him assured eternal life in heaven. We have never seen people so grateful for salvation that they would weep this much and so openly, when coming to Christ. We fell in love with these precious Indian people.

MAFIA ISLAND

Around this time we were invited to the Islamic island of Mafia off the southern coast of Tanzania. There was a brave Pastor who wanted to see the Gospel come to the Island. God had made us bold as lions and we said yes. The best way to get to the island was by air and we made arrangements and flew over the Indian Ocean in a single-engine plane. The large team went by an old ship with our equipment. The Indian Ocean is deep water, but getting closer to the Island, it was too shallow for the large ship to anchor. It had to anchor at two and a half kilometers out. The passengers, including our team, had to get in small flat bottom native boats to be ferried in because of the depth and the coral bottom. One of our team climbed down a rope ladder off the ship, and while attempting to get in the small boat, slipped and fell into the sea. He was still in very deep water and couldn't swim and desperately tried to grab hold of the side of the small boat but lost his passport and his entire luggage along with some of our equipment he was trying to hand down. The small boats still couldn't make it all the way to shore because it was becoming more shallow the closer they got, so the team had to wade the rest of the way on dangerous coral.

We also had a Tanzanian choir of about twenty-five people from the mainland that accompanied them on the ship; in all, about fifty people who made this trip with us. It was wonderful to have them because they were also great prayer warriors. The Island shook in the spirit as we had nightly group prayers for a breakthrough in this place. In Africa we pray aloud together and for hours and sometimes all night, so it can literally shake a building if you have a lot of people. In this spiritually dark place, we agreed with God to make a difference for His name's sake. "The effectual fervent prayer of a righteous man availeth much". James 5:16b.

The government assigned armed guards because there was so much open hostility toward the Gospel. The District Commissioner and Police Chief were Christians and said they would provide personal protection for our team and us. That was fine with us, along with the angels. "Bless the Lord, ye his angels, that excel in strength, that do his commandments, hearkening unto the voice of his word". Psalm 103:20.

The first day we were on Mafia Island we were told about "Kikwato", of whom they said roamed the island. Kikwato was some kind of spirit being who would manifest with the upper part of his body looking like a person, but his legs looked like cow's legs. It reminded us of the Greek mythological creatures. Witchcraft was strong on the island.

The first night we preached drew only a few people, but our loud speakers were so powerful, the message still reached almost everyone. Many of the Muslim men would not allow their wives out of their fenced-in compounds, even for normal activities. They were virtual prisoners in their own homes. Only the house girls (servants) could even go to the market. The second night of the crusade Loren brought me up by his side and said, "The God I serve loves women. My wife doesn't serve Jesus because I make her do so. The God I serve gives women a choice. She doesn't walk behind me; she walks by my side. The God I serve loves women as much as he loves men."

The next night, many women and children came to the crusade. The wives actually slipped out of their compounds to come to the crusade but the Muslim men came looking for their wives and children. They walked right into the crusade grounds and roughly grabbed them and took them home, but quickly, the women and children would come right back. The message of the Gospel of Jesus Christ was irresistible. Some Muslim men sent word through one of our men, Samu, that they would like Loren to talk with the elders of the island. They *said* they wanted to know more about Jesus.

Samu and Loren went to meet them near the marketplace. We didn't realize they were being set up and it was a trap. Loren and Samu sat in a small enclosed area waiting for them to arrive. A large group of men gathered around them and hemmed them in. Loren talked to them about Jesus, but in the middle of his witnessing, Samu whispered, "Loren, they are not serious about hearing the Gospel. They are getting ready to stone us." Abruptly Loren stopped speaking and got up and told them, "You don't have good intentions. Let us through." They elbowed their way through the hostile men. It was only by the hand of God they let them through, but they continued to follow and taunt them. Loren and Samu knew they wanted to do something bad. They walked toward the police station. The men yelled at them in a threatening tone, "We will be at the crusade tonight." We were told this was the first Gospel crusade ever held on this Muslim island.

Loren reported the incident to the District Commissioner and the Police Chief and they doubled up the armed guards for the crusade that night. The antagonists never showed up. We did have a breakthrough on the island and left taped messages of the entire meeting, which, we were told, were played publicly in the shops after we left the Island. The pastors who invited us were pleased with the results of the meeting.

We chartered a large plane to fly out the entire team because getting to the ship was so dangerous. This crusade made the papers in Dar es Salaam declaring "Christianity had made Inroads" into the island.

ATTACKED

Our next stop was Pangani, off the northern coast of Tanzania, right on the Indian Ocean.

It was beautiful and mountainous as well as beach. Pangani had been an ancient holding point for slaves. For centuries, the Arabs had traveled from North Africa down the eastern coast along the Indian Ocean and stopped in ports all along the way. They made inroads into the sub-Saharan coastal countries and took slaves to be transported on ships throughout the world. They also left young men to marry up to four wives and begin propagating Islam by birth. We were able to visit the old slave holding prisons and it made us aware of the cruelty imposed on the slaves and the horrors they went through. It made us want to preach even more about the delivering power of Jesus Christ to set their souls free. We had come to break the yoke for the millions who were still spiritual slaves. Luke 4:18 says, "…to preach deliverance to the captives…".

After settling into our new camp, Loren decided to take the team to the beach for a swim and to relax. It was peaceful as we watched women seining with their nets close to shore and incredibly they brought up lobster. We bought some for lunch, which cost us one dollar each. Our team thought we were crazy to eat them, but we thought they were also crazy because they ate termites. We all laughed and had good camaraderie and fun.

While Loren took the men in the shallow water to play, I stayed on the beach, fully dressed, with our stuff. I was sensitive to the culture of Islam where uncovering yourself as a woman was a taboo.

I was content to sit on the beach and embroider which was a favorite pastime during downtime. The men were having such a good time. I always try to be aware of my surroundings and noticed some men in the far distance quickly walking our way. I immediately sensed danger and began to try to calculate what I could do to get Loren's attention. The roar of the ocean was so loud that they could not hear my call for help. Before I could do anything I was attacked from behind, beaten up, and the assailants took all of our stuff, including our video camera. I could see the thugs running away from me toward the bush. Before I could yell to the men, they had seen what had just happened and Loren commanded them to get out of the water and pursue the attackers. I heard him say "go get them." I had already begun chasing them myself in a reactionary mode. Loren said he yelled for me to stop, but I couldn't hear him and kept running after the attackers. Finally I heard him say, "Celeste, in the name of Jesus, stop." After that, I came to my senses and stopped, falling on the sand. Loren caught up with me and tended to me while the rest of the team sped after the thieves. I was hurt and after assessing my condition we started walking slowly to town to find help. He asked me why I was chasing those thugs. I told him they got my embroidery and I wanted it back.

Loren's son had gotten married on our last trip home and I was embroidering a wedding tapestry for him and his wife with their names and date of marriage. It had taken me months to get close to finishing it and those guys had taken it in my backpack as well as our other valuables.

Someone came along in a vehicle and took us to the small infirmary in town but the foreign doctor had left to go back to

Germany. They didn't have anything in the clinic except one aspirin and one valium to give me. At that point I said I didn't need it, but I think I was so high on adrenaline that I didn't feel much yet, and I put it in my pocket for later.

Three of our men chased the thugs through the thick jungle on narrow paths up the side of the mountain. After a while, they made it to the top of the mountain but couldn't find the thieves anywhere. They stopped and prayed, "Oh God, confuse them and bring them to us." Suddenly, they saw a couple of the thieves slowly walking toward them. Our men jumped in front of them and confronted them. At first they denied they were the thugs, but our men saw they were hot, sweaty, and dirty. It was evident they had been running hard through the jungle. Our team members grabbed them roughly and one of them, a tall, well-conditioned Pastor about 6'3", grabbed one of the thieves and beat him. Samu grabbed the other one and gave him the same treatment. They tied them up and brought them down the mountain and back to our base camp. Someone had an old Land Rover and they stopped by the infirmary and picked us up. Two of our men had the two thugs between them in the backseat.

The night before this incident we had shown our men the video of the Holyfield-Tyson fight where Tyson bit off a piece of Evander Holyfield's ear. This must have inspired Samu. He was so angry with the thugs for beating me he leaned over and bit one of their ears hard. We drove to the old police station in town. When we arrived, the police had already been alerted and came out to meet us. They roughly grabbed the thugs, tied their hands and feet behind their back, and carried them inside the police station as if they were big packages brought from the market. We gave our report and the police came over and threw the guys on the floor at my feet. They began to kick them with their heavy boots and strike them with their police clubs. It was a frightening experience on top of what had already happened on the beach and I prayed these were the right men and not innocent people. We found out later that was just the Police saying "hello" and they were put through the ringer all night. They finally caught a third guy who was involved. They discovered these were notorious criminals who had been stealing and terrorizing the area for a long time so the police were happy to get them. One was a known murderer the police had hunted for years.

I was asked repeatedly to fill out a statement of everything that had happened and the police were adamant that I go over and over the list of what was stolen to make sure I did not leave anything out. They would go thru the list and then make me do it again and again telling me not to leave anything out no matter how small until I even remembered a bottle of water. We had to wait for some time while the authorities sent for the magistrate from Dar es Salaam. Finally, there was a trial and the thugs had to face us in court. All the evidence pointed to them, but the item that sealed their fate was that simple water bottle. The police had questioned me even down to the brand label of water that was in my back pack. It was a brand that was only found in Dar es Salaam where we had previously come from. The police had found it and held it as evidence that these men were indeed the ones that had attacked me and stolen our things.

The five of them had been handing off the stuff to each other as they ran up the mountain and threw the water bottle away as they ran as unimportant. The three who were caught were sentenced to fifteen years in prison. It would be difficult for anyone to survive that long in an African prison. We prayed for those guys and their families that came to court that, even yet, they could be saved. Of the two that escaped, one was later captured and we never heard about the fifth one.

I struggled with knee, neck and a back injuries sustained during this attack, but was determined to keep going and "walk it out", believing God. I would have had to go to Dar es Salaam to the hospital and I didn't want the crusade or the trial to be sacrificed by leaving the area. I saw the five men coming but they were far away. Even then, as I said, I sensed we were going to be attacked. I tried to think how much time it would take to either get to Loren and the team or grab up all the stuff and run toward them in the water. I was thinking of all our money and cameras and all the team's personal items. I had no time to do any of that. It's like they knew when I saw them and just ran to attack me from behind, striking me over and over on the back, shoulders, neck and head. I didn't remember anything except being encased in a bubble of the Holy Spirit. I didn't feel the blows or being dragged in the sand. It was then that Loren and the team saw from the Ocean what was happening. Loren said it was a horrible feeling. They tried to run out of the water but were unable to get anywhere. They struggled to get traction but the waves knocked them down. It took them valuable seconds to get their wits about them and get out but by that time the attackers were almost to the mountain. We thank God for His protection. They could easily have killed me …but God.

We became somewhat celebrities in that Islamic community since the police had captured the criminals who had terrorized the community for so long. It made the people want to come to the crusade to see the brave woman and the men who caught them. Some still threw stones while Loren preached, but none of them hit and we never stopped telling them about Jesus.

CONFERENCES

Over the years we had spent in Tanzania, many had taken advantage of us at every opportunity. We always tried to think the best and at first thought that we didn't understand them because of the language and cultural barriers. Finally, the truth became evident and the Lord began to deal with us about having conferences with pastors to bring them back to solid Bible teaching.

After the Pangani crusade, we loaded the team and our equipment up and went back to Dar es Salaam to hold a conference on the subject of how Christianity was being corrupted and how much of the ministry had gotten off the truths of the Bible and were preaching their own gospel with their own agenda. "But I fear, lest by any means, as the serpent beguiled Eve through his subtlety, so your minds should be corrupted from the simplicity that is in Christ. For if he that cometh preacheth another Jesus, whom we have not preached, or if ye receive another spirit,… or another gospel,…vs 13, for such are false apostles, deceitful workers, transforming themselves into the apostles of Christ." 2 Corinthians 11:3-4 and 13. About a hundred pastors attended this meeting, but it didn't seem we made any progress with persuading them but God was preparing and stretching us into a new aspect of the ministry.

ADVANCED FLIGHT TRAINING

We went back to the states for a few months, continuing to raise money not only for the work in Africa, but for the plane we believed God would bring us. As a pilot, Loren needed more training and enrolled in a flight school in Oklahoma where many pilots go to train to be airline pilots. He needed to become a professional to fly a DC-3. He had struggled in math, so this was especially difficult for him. He would call me every night and say, "This is the hardest thing I have ever done." I encouraged him to focus on what God said about him, "I can do all things thru Christ which strengtheneth me", Philippians 4:13. Then we would pray together. We both knew we were "in school" for more than ourselves. This would affect whether we would be able to win many more people to the Lord. Loren determined in his heart and mind to conquer the math with God's help. Instrument training seemed an especially difficult challenge and means you fly using only the information from your gauges and instrument panel. This prepares a pilot to fly in IFR (instrument flying rules) where you can safely fly a plane in the clouds, at night, or where there is low visibility.

One night Loren called me after an especially difficult day. His IFR flight instructor had stared across the cockpit at him in a not-so- flattering way. He was totally frustrated. He said he looked at his instructor and told him, "I don't care what you think; I know I can do this."

One of his roommates was a pilot in the military and he told him, "They are being harder on you because they know you are going to be flying in Africa; you have to be the best."

After another grueling training flight, his instructor brought him in before the supervisor. They asked Loren, "What do you think of us?"

They thought he would blast them, but Loren answered, "I think you are trying to save my life," which was a true statement.

They were dumbfounded. He finally finished and got his ratings of instrument, commercial and multi-engine, although he still had to get checked out in the DC-3 when we got it. We celebrated his conquering himself and attaining these significant ratings. It meant reaching more people with the Gospel.

By faith, Loren designed a big portable lightweight aluminum platform that so we would be able to carry it on the plane. Although it was expensive, this would keep us from spending so much on buying wood everywhere we went and then just having to leave it. We were in agreement on everything, and I didn't hold him back in his creativity. We both had a bold vision to do something significant in the kingdom of God to reach more people with the love of Christ. We didn't want our lives to be wasted.

A BLESSING

Loren returned to Arkansas to preach for our friends again. We needed a financial miracle to make our next trip back to Africa. While praying on that Sunday afternoon between services, the Lord reminded him of three hundred dollars he still owed on the apartment the flight school had provided. We needed money badly, but knew it was the Spirit of God dealing with him to pay that bill. He had to obey. That Sunday after church he wrote out a check and mailed it that afternoon before he preached again that night.

After the evening service, he walked back to greet people as they were leaving the building. A lady came up to him and said, "Brother Davis, Wait here, please. I have something the Lord spoke to me to give to you for your wife." She came back in a few minutes and handed him a brand new mink jacket. I hadn't been able to go with him because of the expense of the trip. He jokingly told me later he was almost tempted to sell the coat and put the money on the DC-3, but he said he loved me and knew the Lord wanted to give me a special blessing for doing without so many things in order to work for Him in Africa. The offering that day was also one of the largest offerings we had ever received. "He is a rewarder of them that diligently seek Him." Hebrews 11:6b.

Chapter 6

Ashes

W ith God's help we were finally able to purchase a DC-3. It was miraculous how far we had come with such a small base of donors. We found a plane in Franklin, Virginia that was supposed to be in good shape, and Loren took an AP (aircraft) mechanic and checked it and the logbooks out. Although it needed some work, including one new engine, we bought the plane. It needed to have the ailerons, rudder, and elevator recovered with new fabric and Loren ordered this done as well as to order a rebuilt engine that would be mounted on the left side. We were excited.

We planned to use the plane to take our crusade equipment, our team, and ourselves to many cities and villages deep in the interior of Africa. This would greatly speed up the work and increase the amount of souls we could win to the Lord.

FLYING IT HOME

The day came when the plane was ready to fly home. The surfaces had been re-done and the rebuilt engine had been mounted on the plane. Loren had connected with another DC-3 pilot and flew with him back to Franklin to pick up the plane and ferry it home. When they got to Virginia, they were shocked at the weather. A freak ice storm had blown in behind a bad snowstorm. They were aghast to see the plane was covered with about four inches of snow and ice. The runway was in the same shape. Loren was not one to give up easily, but this was a challenge. Not to be defeated, he rented a hot steam pressure washer and cleared the ice off that big plane. The county kindly sent out a grader and cleared off the runway. They set up some scaffolding and put shop heaters on them to warm the engines up enough so they could start them. After starting the left engine, it backfired badly. He said it sounded like cannons went off. They shut it down and the mechanic put new spark plugs in both engines. This time the engines ran smoothly.

They now felt that it was time to fly. There was no heater in the cockpit and it was freezing cold. Loren had taken a heavy orange hunting suit, toboggan hat, and heavy gloves with him for the winter flight and flew as the co-pilot. The captain put the throttles down and the plane lifted off smoothly and headed down the eastern seaboard flying by many naval and air bases. Although it was cold, it was a beautiful flight and they were ecstatic. They landed in a small town in Alabama to spend the night. Loren said he was astonished at how many people pulled off the road to look at the plane. It caught a lot of attention. The DC-3 had been a major player in World War II in the Normandy Invasion, the Burma Lift, and also was used as "Puff", the famous gunships in Vietnam.

The next day they flew into our hometown. I had gathered a few of our friends at our small airport to greet them. It was wonderful to finally get Loren home with our new plane. A few years earlier, while preaching at a church in Oklahoma, we met an AP mechanic, Steve Kirfman and his wife Esther. Steve told Loren at that time, "If you ever get a plane I'll give you an annual on it as a blessing." It was fun to call him and tell him we had a plane and that we were ready for that annual he promised. Steve asked, "What kind of plane did you get?" When we told him it was a DC-3, he nearly fainted, but he was a man of his word.

Steve and Esther came down and stayed with us in our home to work on the plane. Our other AP friend, Alan from California, and my younger son were also AP's and came out to help. Excitement turned into many long hours of hard work. Steve would leave work on Friday evening and drive six hours to our place and work on the plane all weekend every weekend. Esther and I would go out to the airstrip where the team worked and encourage them with words and food. I sensed there was something spiritual about the plane, as if "she knew she would do something special for God." I could feel "waves of love" coming off her and mentioned it often.

One evening after working all day on the plane, we sat eating together at the house, and our phone rang. Loren answered and the man on the other end said "I heard you had a DC-3 whose N number is ***** Is that true?"

Loren said, "Yes."

He continued, "I have been looking for that plane for years. I used to own it. That was the most wonderful plane." He said he used to take electronic goods into Mexico with it and one occasion the federales caught them on the runway about to take off. They opened fire on them. He said, "We didn't have time to shut the door but hit the power full throttle and went barreling down the runway fully loaded." Although the plane was riddled with bullet fire, he said that plane got them out of there alive.

Steve asked "Who's on the phone?" He detected the excitement in Loren's voice.

"He says he used to own the plane," and named the city he said he was from.

Steve was shocked and said "I know him. I used to work for him." Loren handed him the phone and he talked to the man for a while. He invited us to bring the DC-3 to his hangar and work on it there. That was a real answer to prayer. We were all amazed at the phone call and the events of the evening. How he found us we cannot recall.

The man in Virginia we bought the plane from said this plane had flown the famous Burma Hump and supplied our soldiers in WWII, and had been used in the Vietnam War, and rescued Laotian refugees caught in a terrible fire fight. He said that the movie "Air America" portrayed this event, and our plane was the actual plane used in the rescue. On the left side of the fuselage near the tail and on the top of the right wing were many square patches covering bullet holes. You could tell it had been through tough battles.

We felt inspired to name the DC-3, The Chariot of Fire, after the chariot of fire associated with Elijah in the Bible. "…behold, there appeared a chariot of fire, and horses of fire…and Elijah went up by a whirlwind into heaven". 2 Kings 2:11.

Now *this* chariot of fire would take the Gospel to the lost and help rescue untold thousands from spiritual darkness and help get them to heaven.

Loren and another DC-3 pilot flew the plane to our new friend's hangar. They flew low and followed the route right down the middle of a major interstate and again caught a lot of attention.

The old owner was so excited when he saw the 'N' number. It was like he found a long lost love. Immediately he towed the plane into his hangar and began to reminisce about it. The hanger was not far from Steve's home and now, every day after work he was able to go over and work long hours doing the annual and making sure the plane was in top shape. We drove over also and stayed in the area to work on it as well. Loren sanded and prepared the plane for painting which was a huge task. The DC-3 had a ninety-foot wing span, was sixty-five feet long, and with the wheels down was seventeen feet to the top of the cockpit. Sometimes, another friend from church would come and help him, but Loren ended up doing most of the sanding himself. Having been a contractor and involved in industrial painting, he knew what to do. It was truly a labor of love. We preached on the weekends and then returned to the hangar to work on what felt like an eternal vigil during the week.

Finally the day came to actually begin the painting process on the plane. A man from one of the churches we had ministered in donated the primer and the expensive Emron paint. Loren applied the primer on the plane and was in the process of painting the left side. With Emron paint you must paint the whole plane without stopping or the paint will not blend together. About half way thru the process, the airport authorities came over and told him he had to stop painting. He tried to explain, "We can't stop the painting now or the paint job would be ruined. The paint is so expensive and I have spent weeks preparing the plane for this application."

The authorities told him, "We don't care. Stop now, and promise you won't paint again."

Loren was so distressed. Our friend told us "Loren, get out of here now and don't ask me any questions." We left downhearted.

As we walked out to the parking lot, darkness fell. The hangar doors were closed and unknown to us, our friend resumed the painting. He left a lookout to warn him if the authorities returned. As soon as they were spotted, he would stop painting and turn out the lights in the hangar. This cat and mouse game went on for hours. The next day when we came back to the hangar, the plane was painted.

The authorities also returned and when they saw Loren, they spoke sternly, "I thought I told you not to paint this plane."

Loren said, "That's true, sir. I didn't paint it. I stopped when you told me to and we left the hanger. I came back today and found it painted."

The man was about to blow up. There was a major EPA ordinance we didn't know about that forbids spray painting unless there was a certain type of ventilation in the hangar. The man jumped all over the owner asking, "Did you paint this plane?" The owner just shrugged his shoulders and acted innocent. The basic white paint job was done. From that point on, Loren took a special roller that worked with oil-based paint and began to paint the blue trim on the engines and body of the plane. There were no regulations against using a roller and finally, the whole paint job was completed with no further incidents.

A handmade sign was hung on the Pitot tube under the front of the plane's nose. It said, "To Africa with love". The next morning, we were there when the Fire Marshall showed up. He looked at the sign and up at the plane and said, "It sure is beautiful."

He shook his head and said, "Y'all don't do this again." He smiled and walked off.

We had the ministry name lettered professionally on both sides of the plane and had a terrific graphic of horses placed on both sides of the cockpit with the name, *The Chariot of Fire* under it. Researching the scriptures we found, "…fear not: for they that be with us are more than they that be with them. And Elisha prayed, and said, Lord, I pray thee, open his eyes, that he may see. And the Lord opened the eyes of the young man; and he saw: and, behold the mountain was *full of horses and chariots of fire* round about Elisha". 2 Kings 6:17.

THE DEDICATION OF THE DC-3

Excitedly, we sent out invitations for the dedication of the DC-3, The Chariot of Fire. We would soon be taking her to Africa. Partners came from miles around and even from other states to this great celebration. We wanted to formally dedicate this great plane at the small airstrip near our home. We erected the big aluminum platform, 24x32x7 feet tall, to show our partners what it would look like for a crusade. We had celebration flags up everywhere, live Christian music, cold drinks, snacks and of course, good ole' Texas barbeque.

While everyone happily visited with one another, the big bird flew in and wowed the crowd. Under full power, it made a low thunderous pass across the runway. It was big, beautiful, and impressive. When it came by the second time and landed, many people cried. They had all been a part of making this vision come true. What a great tool to be used to spread the Gospel in Africa. We had said over and over again publicly, "this plane will win millions to the Lord." The Bible says, "Without a vision, the people perish". Proverbs 29:18.

The plane taxied to a stop behind the big platform, towering over it. She was seventeen feet from the ground to the top of the cockpit with a ninety-five foot wingspan. There was great rejoicing and shouting that day as everyone took turns going inside and inspecting it. Truly, this was the dream that had come to pass for us. My parents had already gone on to be with the Lord, but our children and Loren's eighty-five year old mother, were with us that day. She shared from the platform what this plane would do in the kingdom of God. One partner who was a musician wrote a song in honor of our vision and called it, "The Chariot of Fire." He gave us permission to use it and I recorded it and have sung it in many nations of the world.

THE UNTHINKABLE

The next week Loren began intensive training in the DC-3. The first step was to learn how to taxi that big plane. He enjoyed explaining it to me. He said it was squirrelly to handle on the ground because it had a tail wheel instead of one under the nose. You have to control this big tail dragger with differential power, which is using opposite throttles to turn it. After the plane gained enough speed, it could then be controlled by the rudder, which was controlled by the foot pedals.

The day before catastrophe happened, we went out to the plane to check on her and to pray over it. I immediately sensed something was very wrong. My spirit was deeply disturbed. Loren told me, "The FAA had three inspectors out here this week checking the plane out and totally cleared it. The plane is airworthy".

I said, "I don't care, something is wrong." We inspected the doors and Loren assured me, there was no change on the doors, but my spirit was very upset and I didn't want him to fly the next day.

He reminded me, "My instructors are coming in the morning and I have been working for some weeks on getting my type rating for the DC-3. The plane is okay, don't worry." The next morning, I was still very upset and again asked him not to fly. Loren was too excited and insisted he was going to fly so I drove him out to the airstrip where we were to meet his instructor at the plane. When he arrived, he didn't look happy. He told us he and his wife had a bad argument that morning. One of the things pilots are taught in flight school is never fly a plane if you are upset or are having a domestic problem, but the excitement of flying this big bird caused both of them to ignore this. The Captain told Loren to get in the captain's seat. He would be flying co-pilot and watching Loren. He instructed him to fly to an airport about fifty miles away because it had a wide and long runway. This would assist him since it would be his first time to land the DC-3 as captain.

Loren pushed the throttles slowly on those two big twelve hundred-horsepower engines and began to move down the runway. He said it was exhilarating having his hand on that much power. They flew to their destination and Loren set up the approach lining up on the runway. He had a perfect landing with no bounces. His instructor and Loren both knew he would be able to handle this big bird. They taxied over to the main FBO and went inside to meet the other instructor who would work with them. The second pilot needed three takeoffs and landings to get current in the DC-3, so Loren told the pilots to go ahead and fly. He would wait for them in the lounge and get a cup of coffee and would catch up with them when they had done their touch and go's. The second instructor said, "No Loren. Go with us. You need to watch how we work together." That made sense, so he agreed.

Loren told me later he would never forget how beautiful the plane looked as he walked toward it. It was so big and marvelous. He admired her and what it would mean to the ministry in Africa. Little did he know this was the last time he would see her intact. He had the plane filled up to its eight hundred gallon capacity of gas. The two pilots got in ahead of him and sat down in the pilot seats. Loren sat behind them in the cabin. The engines started with a great rumble but those big engines sounded like a great symphony to Loren. A big smile was all over his face and he praised the Lord for the great things He had done to get us to this place of outreach to the lost.

They had barely lifted off the ground and the co-pilot motioned for Loren to join them in the cockpit. Hardly had he gotten up there when the captain said, "Something is wrong with the left engine." That took the smile off Loren's face. The airspeed was too low to keep the plane flying, so the captain had to put the nose down to keep the plane from stalling. Loren stood between the two pilots and couldn't believe what he saw out the left window. The trees were not far below them and they came down closer and closer to them. It was then it hit Loren they would crash.

He stood in the cockpit, not in a seat, had no seatbelt on and no time to get back to his cabin seat. Things happened so fast, the only thing he could think of doing was to squat down behind the captain's seat and wait for the inevitable. Someone asked him, "What were you doing? Were you screaming, 'We're going to die'?" The truth is, he said had no thoughts of death. He said he had no fear. The only thing he thought was, *Oh no, our plane!* Then the wings hit the tops of pine trees. They went down in a forest in a residential area. He told me he saw a woman mowing her front yard. When she saw the big plane coming down, she dropped her lawn mower and ran to grab her young daughter off the porch and fell on the ground to protect her. The plane barely hit the top of her house, a mobile home. Thank God they were not injured.

They were one hundred and fifty feet from smashing into another house across the street. We later discovered there were a mother and five children in that house. He said it felt like they were all going to die. This big plane was loaded with eight hundred gallons of gas and weighed twenty-two thousand pounds. It was virtually a flying bomb. They hit the ground at 110 miles per hour. Any way you look at it, with the law of physics, it takes a long time to stop that much weight going that fast. In the natural, there was no chance, but God...

After tapping the roof of the mobile home, the plane went perfectly between two oak trees. They were the exact width of the fuselage. The oak trees were about eighteen inches thick and gave a little before they snapped. Instantly it tore off the left wing. The nose of the plane then rose to plow into the home and everyone there, but something happened. The tail caught between the two oak trees and stopped the twenty-two thousand-pound plane going 110 miles per hour instantly like it had landed on an aircraft carrier. It's amazing it didn't completely tear the elevator off the tail. The captain's head slammed into the windshield. The co-pilot's head went through the double plated tempered glass windshield. It should have cut his head off. Both of them should have been killed instantly, but it didn't even knock them out.

Loren was hunkered down behind the captain's seat and when the plane crashed he said it felt like someone "moderately slapped my back. It didn't even knock the breath out of me". Even though he wasn't in a seat and had no seatbelt on, he was not even thrown around the plane. I said he must have been sitting in the lap of a big fluffy angel. As he sat on the floor he said he thought, *Is this all there is to a plane crash?* Within seconds, he had a big attitude adjustment. Fire came down on his head from the electrical bus box behind him. When the fire hit the top of his head, it got his undivided attention. He said he rolled over toward the pilots on his hands and knees and saw the plane was on fire. The co-pilot was stuck in his seat with fire all around him and said later he thought he would "fry". The skin of the fuselage on the right side of the plane was torn off and

Loren said you could see the yard outside. The captain fought through the flames and rescued his friend. That was a real act of love and bravery.

They both turned their heads toward the rear of the cockpit and saw Loren on his hands and knees in the aisle behind them with blood streaming down his face. They were shocked, fully expecting he would be dead. The captain yelled, "Fire, get out!"

Loren told me he didn't have to repeat that command. All this must have been seconds after the crash. The eight hundred-gallon fuel tank was situated right under them and they knew this big bird was ready to blow. They got up and all of them ran as hard as they could to the doors in the back. They tried to get out of the big cargo door, but, it was twisted and wouldn't open. Fire chased them down the aisle, even singeing the back of the captain's hair. With the cargo door jammed, there was only one escape left, a small passenger door situated behind the cargo door that opens by lifting upward and bracing it open with a latch. When they got to the small door, knowing it was their last and only hope, they couldn't believe what they saw. The door was open and lifted upward without anything holding it open. For sure, they knew something supernatural happened. They ran out the door as fast as they could and collapsed in a ditch twenty feet to the side of the tail. Loren said he ran so fast he was sure in that moment he would have won the one hundred meter in the Olympics.

THE FOURTH MAN

The plane had come to an abrupt stop right under some heavy power lines. Thank God they didn't hit them. All three men barely got out of the plane when they heard three explosions and then a massive boom. Fire shot up higher than the trees. It was so hot, Loren tried to get up to run again, but to his dismay, he found he couldn't move. He felt like the heat would kill him and wondered how he had been able to run out of the plane, but now couldn't move. He cried out, "Somebody help me. I can't stand the heat." The co-pilot came over and laid his body on top of Loren's body and took the heat for him trying to shield him. A neighbor brought a pickup truck and parked it between them and the inferno to help keep the heat away from them.

Loren said the co-pilot told him there was a fourth man in the plane. Loren said suddenly it felt like his insides went through a meat grinder. To his astonishment, he saw something like a ghost starting to come out of him. This transparent figure pushed out about four inches through his body. It looked like there was another person inside him who tried to come out. He said he couldn't explain how he saw this, but it was then that he realized he was dying. For sure he was in physical trauma after the horrific impact of the crash and also in emotional trauma having lost our plane, our tool for the vision for reaching the lost in Africa. Then the Holy Spirit spoke to his spirit, "This plane is not your ministry. You still have your wife and children. You still have my anointing on your life. I'm giving you a choice. You can live or go home?" Loren had so many great disappointments in his life in the past, but something inside of him wouldn't let him give up. He said he bit down on his lip and said with great determination, "I will not die over this plane." As he spoke, he said "my spirit came back into my body". Then it came out again. He said this happened three times, seeing his spirit seesaw back and forth, and three times he said the same thing, "I will not die over this plane." Then finally his spirit settled back in his body and calmed down.

Later, the co-pilot who had used his body as a shield protecting him from the heat told him, "Loren, I fought in Vietnam. I've seen many men die. Loren, you were dying."

Loren acknowledged him, "I know it."

Those big propellers, twelve feet in diameter and weighing eight hundred pounds each, flew off the plane when they hit the ground. Still spinning one of them catapulted and hit a window air conditioner with one of the blades. It obliterated the air conditioner, and then strangely bounced off. The other giant propeller tore through the front porch overhang of the house like it was butter, but then within four feet, immediately came to a stop, rolled over, and broke the window to the living room. Then it softly settled one of its blades on the back of a couch next to the window. Later we were told there was a nine-year-old boy on the couch and he wasn't even scratched. As big as those propellers were and as fast as they spun, it's nothing less than a miracle they didn't completely go through the house. The twenty-two thousand pound plane going 110 miles per hour stopped within one hundred and fifty feet, merely twenty feet from the house. That defies the laws of physics.

Ambulances came with body bags expecting to pick up human remains. One of the ambulance drivers asked the captain, "Where are the bodies?"

He said, "There are no bodies. I was in the plane myself."

News helicopters and media appeared to show up immediately out of nowhere. Some FAA officials were overheard later talking about the crash. They said there was not a chance anyone could have survived. But God...

I had watched them take off and then had driven over to the local Christian bookstore owned by a friend of ours. I was still troubled in my spirit, but had no choice but to leave them in the hands of the Lord. I was in the back office for about half an hour working on some ministry business when I became ill. Suddenly I felt sick and doubled up over the desk. I couldn't imagine what happened. Every muscle in my body ached and I felt helpless. I started to pray in the Spirit and then realized I was in intercession for someone else. I kept praying. This went on for about half an hour and that's when someone burst in and I heard the words, "The DC-3 crashed; the DC-3 crashed." It was already being reported over the television and radio stations in special breaking news alerts.

My head spun in disbelief as I walked into the store area to confront the person shouting those words. I could not grasp in my mind that it was "our" DC-3. After short questioning, I realized it *was* ours. It was then that I knew I had interceded in prayer for the crew and Loren. I also "knew" in my spirit they were not dead. Stopping long enough to call the hospital, I was able to get through to ask the nurse about everyone's condition and let them know I was on my way. The nurse confirmed everyone was alive.

Immediately, a pastor friend volunteered to drive me and we began the hour drive to the hospital. The entire trip a scripture ran over and over in my mind. Jesus said, "Except a corn of wheat fall to the ground and die it abideth alone". John 12:24. *God,* I thought, *you can't mean this plane is a seed.* It went against everything we had been taught about God.

"Lord, do you mean you wanted that plane to go down, you wanted this thing to happen?"

The Holy Spirit spoke plainly to my spirit, "No, I am not the devourer, but 'all things work together for good to them that love me and are the called according to my purpose." Romans 8:28. "Trust me".

Thank God, instead of going to the morgue, we were headed for the hospital. The two pilots came into the emergency room where Loren was taken and the co-pilot said, "Loren, if you wouldn't have been in the plane we would have died. There was a fourth man in the plane."

The two of them were released and did not even have to spend the night at the hospital. Loren was the most banged up. He had a bad wound on his scalp and some cracked ribs. His blood pressure was up and he was in shock. They wanted to watch him for the night. No one was killed or burned on the ground. Our boys rushed immediately to the hospital. One of them drove from the other side of Houston when he heard about the plane crash on the radio. He was terrified and rushed to the hospital as soon as he could. Loren's elderly mother was in a nursing home and unfortunately saw the accident on the news. Loren called her immediately and told her he was alive.

His sister-in-law drove down the highway and saw a big ball of smoke coming up from the crash, but of course, didn't know it was our plane.

One of our boys said, "If anyone ever went to the crash site and saw what was left of the plane and where it landed and everybody came out alive, they would believe in miracles."

Loren had a big gash on his head and they had an intern giving him shots in his scalp and clumsily tried to sew him up. It was excruciatingly painful. I found him in a lot of pain, not only from the blow to his head and the cracked ribs, but the horrific emotional trauma.

I tried to comfort and encourage Loren and myself, reminding him that what looked like "evil... but God meant it unto good." Genesis 50:20. Somehow, God would turn this around and help us understand.

About two hundred of our friends showed up at the hospital in a wonderful show of support. So many had helped in the vision and losing the plane was a blow to them too.

After being monitored all night, Loren was dismissed. One of the kids had brought fresh clothes; oversized shorts and t-shirt for Loren so he could leave the hospital. He wasn't able to wear shoes because his feet were so swollen. He looked like he had come out of combat.

Neither of us had slept much and we went from the hospital directly to the crash site. Esther and Steve had driven down and went with us. Naturally he was interested because he had invested so much time in this project and dream.

IN THE MIDST OF THE ASHES

Walking through the ashes of our plane, our dream was an unspeakable nightmare. We were both in shock. The fire had been so hot it incinerated most of the fuselage. Only the two burnt out engines and the tail section were left. While looking through the carnage of what used to be the cockpit, we found the thick International Aviation Directory in the midst of the ashes. This book shows all the airports in the world, their coordinates, and services. The cover and some of the pages of the book had been burned off completely but there remained a few singed pages. It was amazing anything was left because the fire had been so hot. The book was opened to "Kenya." I said, "Look at this Loren," showing him the displayed page. We had been sensing that our work in Tanzania was coming to an end as far as a base was concerned. Could it be a sign that God had special work for us to do in Kenya?

Loren said, "It's amazing the book didn't completely burn up, but I don't believe this is a sign that God wants us to go to Kenya." We knew there were many missionaries in Kenya, and had up to this point, no thought that we were needed there. We did not know this *was* a hint of a major work God had planned.

The crash scene was painful for all of us to witness. In Loren's beat up condition we walked with Steve and Esther to see the owners of the houses that were damaged. We wanted to assess the damage done to their properties and the people and assured them all we would make everything right and take care of them. We spoke out of our hearts. The law did not require insurance since it was not a commercial airplane. We never dreamed of this scenario and now, we had lost our major ministry asset.

We spoke to the lady of the house where the plane had crashed in her front yard. She said, "I have just one question to ask you. Who was that man? After the crash, your captain went over to our faucet and washed the blood off his face. A stranger showed up from the woods behind the house. He was a man with a dark beard and was dressed in some nice casual clothes. My neighbor saw him too, but we didn't recognize him. He wasn't from around here. He followed the captain around the yard. Then he came up to me and asked, 'Are you all right?' I said, 'Yes,' and then he walked toward you while you were still in the ditch before the ambulance came. I turned my head for a second to check on the kids, and then looked back. When I did, I didn't see the man. He had disappeared."

Then she asked a pointed question, "Who was that man?"

We remembered at the hospital the co-pilot told us there was a "fourth man" in the plane. We knew there were only three in the plane. This didn't make any sense. The co-pilot had said "Loren, if you wouldn't have been on the plane, 'the fourth man' wouldn't have been here and we would have all died." This reminded us of the story in the Bible from Daniel 3, when a "fourth man" showed up in the fire with Shadrach, Meshach, and Abednego and delivered them. The fourth man was the Son of God. *"Could this have happened again? Was the Lord with them in the crash? Yes, we could believe that."*

The pilot got a lot of praise for flying the plane right between those two trees which stopped the plane, but later confided that things happened so fast, he never saw those trees. Without a doubt the Holy Spirit guided that plane through that tiny space.

THE NEWS MEDIA

We weren't the only ones at the crash scene that day. The news media was all over the place. Some of the neighbors were stirred up against us and made a big scene in front of the media. Of course the press and TV loved it. The photo and story of the crash was on the front page of the *Houston Chronicle*, on CNN International, and many other national and international news media. The owner of the house where we crashed stood in his front yard with a beer can in his hand and was in a rant against us. He didn't thank God for saving his family and home, but attacked us for crashing in his front yard.

The real emotional impact of this whole thing didn't come down on Loren until we got home later that day. His ribs were cracked and the only way he could get any relief was to lie down on the couch. It was extremely painful and difficult for him to get up and down. He had to slowly roll off the couch onto his knees. We really appreciated a pastor friend and his wife who came to stay with us in our home to comfort us and help in any way they could. The second night after the crash, about midnight, the emotional trauma unloaded. Loren began to sob and cry uncontrollably. We held each other sitting in the middle of the floor and I cried with him. It felt like our hearts would burst because of the grief and sorrow. The news media from everywhere constantly called the house, but I intercepted the calls and shielded Loren from them. Legally we were in a precarious situation.

Some partner churches sent teams of carpenters and other workers to repair the houses and clean up damage done in the neighborhood where the crash occurred. We will never forget their labor of love. It was expensive to remove the wreckage and clean up the soil because of the oil spill. The Lord helped us to have all damages repaired within a week, yet the neighbors and news media were still incensed against us. Of course, outsiders and lawyers were agitating them.

One neighbor told the news media he saw fire come out of one engine before the plane came down. After talking with our AP mechanic about this, he told us there was nothing wrong with the engine. The reason they saw fire coming out of the bottom of the engine was because the captain had jammed the throttles forward. This caused the fuel to bypass the intake and go straight to the hot exhaust pipe, which caused fire to shoot out. The owner of the house where we crashed threatened to sue us even though we had repaired his house and property.

DIVINE INTERVENTION

Loren still suffered in body, mind, and spirit and had a hard time pulling himself together. He wanted to see his daughter who lived out of state so I immediately packed him in the car and drove us to visit her. We prayed a lot along the way and it did our hearts good to see his daughter and grandchildren.

About three weeks had gone by and we had returned home. We prayed about this impending lawsuit and asked God to intervene. We had lost our major ministry asset and didn't know what a lawsuit would do to us.

After praying specifically about this, we got a phone call from the homeowner. He said, "Loren, I'm ready to settle. Can you come over to see me?"

We told him, "We will be right there."

When we got to his home an hour later, he met us outside. He didn't look like the same man. He was clean-shaven and friendly and invited us into his house.

He said, "Let me tell you why I'm dropping this suit against you. Not long after the crash lawyers had contacted me from all over the country about suing you and I had decided to sue you, but from that day forward, my world turned upside down. I literally went crazy and ended up in a psychiatric ward. While in there, I began to die. I saw God and He asked me, 'Do you want to live?' I answered, 'Yes, God.'"

He said God had shown him hell and told him, "This is where you are headed."

"This is your last chance. Either you straighten your life out and get right with me now and also leave Loren Davis alone or this is it."

The man told us to "write down anything you want. Don't give me any money. I will sign whatever you write and release you."

Loren wrote up the release and they both signed it. As a result, he got saved right then and there. A few weeks later, he joined Loren on a TV program and gave his testimony. He told the audience, "If it wasn't for this plane crash in my yard, I wouldn't have found God."

The Lord gave us a scripture promise at this time. It said, "I would like you to understand brethren that the things that have befallen me have turned to the furtherance of the Gospel". Philippians 1:12. The news media came for more interviews and now gave Loren a large platform to testify of the miracle power of God that caused him to escape from this burning inferno without any burns. The news media reported that our ministry in Africa was finished, but we didn't believe it. It doesn't matter what others believe about you; what matters is what God says about you and what you believe about yourself.

BACKLASH

A by-product of this crash was that two-thirds of our donors stopped supporting us. One person wrote and said the reason the plane crashed was because we named it The Chariot of Fire. They said the crash happened because the name was a "bad confession" which brought about the fire.

One pastor who had a large television ministry sent us a big bouquet of flowers addressed to Dr. Loren Davis. Since he no longer had anything to do with us, we decided the flowers were meant as a funeral bouquet. Loren was not a doctor so calling him doctor was nothing but satire. Later we found out from members of his congregation that he had taken an offering in his church in our name, but were told by some of his members he used the money instead on his church parking lot. We learned to pity the poor saints of God who love Jesus Christ and are trapped under these perverse ministries. This was our "welcome to the real world" again, and it hurt us more than the crash itself, but it didn't break us. God began to put even more fight in our spirits to continue to believe Him for everything concerning us. "Beloved, think it not strange concerning the fiery trial which is to try you, as though some strange thing happened to you". 1Peter 4:12.

They say you know who your friends are when the chips are down. We learned this can be a bitter lesson. Now, it was almost like starting our ministry all over again. People said a lot of cutting things and it was like pouring salt in our wounds. Some people never consider there is a devil trying to stop the work of God. Although, we have many true friends, we also discovered many people actually enjoy seeing you suffer and fail. Some people thought we were not walking in faith or this would not have happened. If all suffering is caused by the lack of faith, then Jesus and Paul must have been the worst of sinners. We became annoyed at the perverse teachings flooding Christianity and the consequences they had on God's people.

After the FAA investigated the crash, they found there was no mechanical problem. The crash *was* caused by pilot error. During the investigation, the FAA was impressed with our AP mechanic and the job he had done.

WE DIDN'T GIVE UP

When something bad happens to you, it will either shut you down or make you stronger. You are the one who decides. Proverbs 24:16 says, "For a just man falleth seven times, and riseth up again". Much of the news media said our mission to Africa was finished, but six weeks after the plane crash, we went back to Africa. A television reporter met us at the airport before our departure wanting to do one last piece on us before we left the country. I reminded him, "This was just a bump in the road for us." We had to have tremendous tenacity.

On this trip we didn't plan on doing any major crusades; we wanted to prove to the donors we had left, and to our African team, the devil, to God, and even to ourselves that we weren't finished. We weren't giving up. We agreed we would be faithful to what He had called us to do no matter what we had to go through. If we no longer had a plane, then we would take a car. If we couldn't find a car, we would ride a donkey; and if we couldn't find a donkey, we would use our two legs and walk. We would use what we had until we had what we needed. We started out that way and weren't too proud to have to do it again.

The Tanzanian Civil Aviation Authority had been cold to us about basing our plane in Tanzania. Even though we didn't have a plane now, we began to pray and ask God where He wanted us to go. We still had not given up our vision. Rwanda, Burundi, and Uganda were the only positive possibilities on the horizon and all of them had invited us to come to their nations. Now that the Chariot of Fire had crashed, we believed God was somehow going to move again. One thing was clear thru this; after eight years in Tanzania God was leading us in another direction for a base. Our job was to keep working and let God reveal the time and place.

We flew commercially to Kampala, Uganda where we had been invited to speak at a minister's conference and after that went to Kenya to preach for another pastor. We were happy to have one of our boys, and another young man come along with us and our AP mechanic Steve, and his wife Esther. It was comforting to have the fellowship of a small group travelling with us because we were still so traumatized over the recent events. The trip had been planned before the crash and we decided to keep our commitments in hopes that the Lord would use our testimony and help us heal. Kenya turned out to be a turning point in our ministry although we later learned sometimes when God opens a door; there are still many mine fields to get through.

A FRESH VISION

On this short visit to Kenya God put it in our hearts to explore what the Kenya bush was like. We didn't know much about Kenya, but had been told it had a high Christian population and many missionaries. We wanted to be where people needed the Gospel and to go where no one else would go. We drove five hours out of Nairobi and visited the primitive Pokot tribe. They were very unfriendly and not open to outsiders. This was a life altering trip for us. From this visit God birthed in our hearts to go to the interior and penetrate the primitive tribes in Africa. We began to understand that the interior of most of these countries had barely been reached. The Pokot were so fierce, even their own countrymen, Kenyans, dared not enter their territory. Their land was a barren and remote wilderness and the people were wild and violent to the outsider.

After visiting the Pokot, we went back to Nairobi and Loren made a visit to the Civil Aviation Authorities about their position on basing a plane in Kenya. We got a favorable response, although we didn't have a plane yet but, The Bible says, "Faith is the evidence of things not seen." Hebrews 11:1. We didn't know how God would provide, but in our hearts we believed another plane was coming. This was the beginning of Kenya opening up to us for ministry although it would be two more years before we made the permanent change of base.

A MIRACLE

We didn't know that the day after we left for Africa after the DC-3 crash, a man sent a very large donation to us toward another plane. God always amazes us. At this time there were no emails or cell phones in Africa and we didn't find out about this gift until we got home. We hadn't mentioned to our partners yet that we had begun to believe God for another plane so this act of giving was so encouraging. God was already ahead of us before we even asked.

Ephesians 3:20-21 says, "Now unto Him that is able to do exceeding abundantly above all that we ask or think, according to the power that worketh in us, unto him be glory".

Chapter 7

Tents

Less than a year later God had moved through His people and we were able to buy a wonderful six passenger single-engine Cessna 206 which had only a few hours on it. We didn't want to spend years fundraising and God had moved for us to afford this plane. We took what was in our hand. The 206 was much smaller than the DC-3 of course, but it was a top bush plane and would greatly help. It was powerful and ideal for the bush of Africa. Steve helped us once again, outfitting it with a cargo pod, STOL kit, to enhance our short field take- off capabilities on bush strips, and he also set up the rigging on it to pull a banner behind the plane to advertise our crusades. This was a unique thing for Africa and was sure to catch a lot of attention and help bring more people out to hear the gospel. The Bible says, "The righteous are as bold as a lion." Proverbs 28:1. We began to use the plane in the states as we went around sharing the vision. This enabled Loren to build up time in it and become well acquainted with the plane before we took it to Africa. It was pure joy to him to be back in the saddle flying again.

One of our partners wrote a new song for me called, "The Word on the Wing" about our new C-206. The last phrase in the song says, "Burned out dreams can't warm your soul, but what the alpha once begins He will finish in the end; there's a word on the wing." This encouraged us so much and I began to sing that new song everywhere we went.

The Lord had given us a scripture from Isaiah 61:3: "To appoint unto them that mourn in Zion, to give unto them beauty for ashes, the oil of joy for mourning, the garment of praise for the spirit of heaviness; that they might be called trees of righteousness, the planting of the Lord, that he might be glorified."

God's promise to us was that He would give us "beauty for ashes". We were remembering the ashes of the DC-3 we had walked through after her crash. We felt our mourning was over. We called the Cessna 206, "Chariot II" and had "Beauty for Ashes" painted underneath the horses on both sides of the cowling just like it had been on the DC-3. The Lord was giving us joy.

THE CHARIOT OF FIRE II

We talked to a pilot from Zambia about ferrying our plane from the states to Kenya. We weren't experienced at taking on a trip like that. The Zambian said he couldn't do it, but gave us the name of another pilot who might. This man's name was Jim Bone from Chicago. He and his wife, Lonnie, were both pilots and were fine Christians. We contacted them and in time they agreed to fly our plane to Africa. He was a 747-400 captain with a major Airline. Flying our plane overseas would be their twenty-fifth flight ferrying airplanes for missionaries and this would be their last missionary ferry flight before retirement. We were so honored to meet them and that the Lord had provided for our need with such great and loving servants of God.

We had another big dedication service but this time for The Chariot of Fire II; once again at our small hometown airport. So many of our supporters came out again to celebrate and rejoice with us. What another fantastic day. After the wonderful dedication service, we all laid hands on the plane and prayed for it and the pilots to have a safe flight. Jim and Lonnie cranked the plane up and we all waved goodbye, misty eyed, as it headed off to Africa. Believe me, there were many prayers going with them. We finished our commitments in the states and planned to meet them in Nairobi in two weeks.

They first flew the plane to Chicago and then to Bangor, Maine where they installed an eighty-five gallon ferrying fuel tank. The backseats were taken out and the tank was installed. Engineers connected the auxiliary tank to the fuel system from there. At Bangor, they continued the trip to St. John's, Nova Scotia. After that, it was eleven hours straight across the North Atlantic to the Azores. Jim told us later that Lonnie was worried about the engine stopping; she might be eaten by sharks if they went down. Jim told her not to worry about that, they would freeze to death within five minutes, once they hit the water. This was "husband pilot" comforting his "wife pilot" talk.

After refueling in the Azores, they flew another eight-hour leg to Portugal and then down the middle of the Mediterranean to Palermo, Italy. From Palermo, it was direct to Luxor, Egypt; then down the middle of the Red Sea past Saudi Arabia to Djibouti. At Djibouti, they headed across Ethiopia toward Kenya until they reached Nairobi. Jim told us at one point while flying across Ethiopia, the engine started to sputter and he thought they would have to put it down somewhere, but the engine smoothed out. Without a doubt, the Lord helped them make this dangerous trek.

We met them thirteen days after they left our hometown in the states and landed at Wilson Airport in Nairobi. We hugged and had a big celebration because they arrived safely intact. Our ministry had paid for all the fuel for the flight and for the navigation and landing fees along the way. However, they insisted on paying for their own personal expenses incurred along the way. We had determined to be a blessing to them. We put them up and took them to dinner at one of the oldest hotels in Nairobi and we loved the atmosphere there, a garden in the city, and the food was first class. It was also the first stop for us when we first arrived in Africa years earlier and was a sentimental memory for us. After our celebration dinner that night, they insisted on blessing us with an offering for the ministry. We were deeply humbled. It was awesome to meet such a great couple so dedicated to the work of the Lord. Months later, Lonnie and Jim sent us the thirteen day log they kept of the trip flying over "the pond", the name pilots gave the Atlantic Ocean. Jim has since gone on to be with the Lord. We will never forget him and Lonnie and their kindness toward this ministry and to us.

FIRST FLIGHT

We had the plane serviced at Wilson airport, Nairobi and readied it for our first flight in Africa. The owner of a farm up in the Rift Valley was kind enough to let us use his private airstrip. Later he even allowed us to build a hangar alongside his to keep the plane there. We prepared to fly from Nairobi to this airstrip, which was a forty-five minute flight away. An African pastor friend agreed to go with us on the flight to help us locate the farm and airstrip from the air.

This was to be a monumental flight. Our first flight in Africa and our African friend had never been in a plane before. By the time we had all our clearances and were ready to takeoff, it was five P.M. This gave us a short window to fly since the sun goes down around 6:30 P.M. at that time of year, but the locals at the airport assured us it was a short flight and we would make it there before sunset. Loren had no coordinates for the farm, so after takeoff he kept asking 'do you recognize the road below?" Imagine; this African Pastor who had never flown before was so frightened; it was not easy for him to recognize anything from the air. Things looked different from that perspective. He was married and had a newborn baby and later told us all he could think of was, "I'm going to die and I will never see my wife and child again." What a fine aviation team we all made. I sat in the back praying. That was the first of many prayers I would pray as we began this new life of bush flying.

We finally located the farm and landed as the sun went down. The airstrip was grass and of course had no lights. God sure looked out for us. As the plane sat down on the airstrip, we skidded a bit because it was muddy, something that wasn't noticeable from the air, but praise God, we made it. We were picked up at the farm by our host pastor and driven to his large church. We preached there that night and had a big celebration that we and our plane had arrived safely.

LAKE VICTORIA

One of the first trips we made was to western Kenya along the eastern coast of Lake Victoria. The plane gave us the advantage of being able to work much of East Africa quickly which included six countries. Routes that would have taken us days, or even weeks by road were now easily accessible by air. We put together a new team and met them in Mwanza, Tanzania, which is at the southernmost end of Lake Victoria. After greeting them, we flew across the southern part of the Lake to Ukerewe Island. This area was full of crocodiles. The team went by vehicle and crossed over on a small peninsula from the mainland to the Island to rendezvous with us there. We spotted the primitive airstrip and landed successfully, but unexpectedly, the people rushed out onto the airstrip to meet us. The wonder of the plane excited them and Loren had to immediately kill the engine to keep someone from walking or running into the spinning propeller.

The people of Ukerewe were animists who worshipped crocodiles, snakes, and almost everything else. One of their practices was to kill a bull when someone died as they believed this would release the spirit of the dead person to enable him to pass over into the next world more peacefully. If a husband died, after the funeral, the widow was expected to go into the forest and have relations with another man. This was a part of their customs, though hard to comprehend.

We began to pray earnestly that the power of the Gospel would reach these pagan people, and they would open their hearts to believe on Jesus. We were put up in a secure fenced compound on the island. The Lord blessed and many turned to Christ. It wasn't until after we flew back to Mwanza that our new advance man told us we were the first white missionaries to ever survive preaching in Ukerewe. Astonished, Loren said to him, "Why didn't you tell us before we went." He told Loren, "I was afraid you wouldn't go." We learned first-hand that sometimes you can know so much, you are afraid to do anything for God.

BOILS LIKE JOB

Early on in our ministry in Africa, we began to break out with huge carbuncle-type boils all over our bodies. They would never come to a head so we could have them lanced. They would come up in our eyes, tailbones, inside our noses, and move around our bodies. We were in and out of clinics but nothing could help us. This horrible torment lasted on some part of our body without let up for almost five years. It must have been something similar to what Job experienced, yet we continued our work and tried to live the best we could through pain and suffering. I remember being at the altar one night sobbing and crying out to the Lord for deliverance from this awful oppression. They were so painful. I could hardly walk because one had come up on the inside of my leg on my right thigh.

MALARIA ATTACK

We flew to the town of Mwanza on the mainland side of the lake. About three days into the crusade there, high fevers hit Loren. There was a bad boil on his leg and I thought that was causing the high fevers. I dressed it regularly, gave him Ibuprofen and took care of him.

I knew open wounds or sores can go septic fast. One of our first experiences with this was when I got a cut on my foot on our first trip to Tanzania. I wore open sandals and stubbed my toe. Within hours, it had become inflamed and I could barely walk. Fortunately, we were near an American missionary doctor who immediately treated my foot by soaking it in an antiseptic and putting me on antibiotics. The doctor warned that, otherwise, it would be easy for me to lose my foot. From that point on, I was extremely serious about keeping us from any life-threatening situations, no matter how small they appeared to be at first.

At night, the fevers were so high that by morning, Loren's t-shirt was soaking wet and the bed was drenched from sweating so badly, but the fever was gone so we thought the worst was over. We lived in bad conditions and there were millions of mosquitoes. It was hot and humid, and malaria bred easily around Lake Victoria but we had not connected the dots. He was such a strong man of faith and never complained or entertained any thoughts of illness.

In the daytime, his wound would feel better so he would go ahead and preach at night. God moved in this meeting, however, every evening after the crusade when we returned to our room, the high fevers returned and we spent the night praying again.

The last day of the crusade it was obvious he was very sick. We stayed in the car behind the platform until it was his time to preach and slowly went up the back steps to the platform. He could hardly move. By now, the fever was out of control. The Lord helped him preach, but afterward, I took him to our room along with the interpreter who was also a medical doctor. He realized Loren probably had malaria and took a quick blood test and ran it back to the lab. The test was positive. Mwanza region is known for having a deadly strain of cerebral malaria that kills hundreds every year. In fact, at that time, malaria was the number one killer in Africa with HIV-AIDS coming in second.

We took him to the local city hospital for admission and treatment. It was humiliating to Loren to have to be taken to this Muslim hospital after preaching Christ for salvation and praying for the sick, but at this point, we just had to trust the

Lord to deliver him in whatever fashion He chose. We needed help. There was no choice. Our Christian doctor had to go out of town, but he had an Indian doctor friend working at the hospital who was also a Christian and he turned us over to him. The hospital policy was that all the workers at the Hospital who were not Muslim had to sign a contract they wouldn't witness about Jesus Christ or pray in His name while on duty.

Loren was delirious by this time. I stayed with him day and night hovering over him praying. You have to have your family or someone to help you while you are in any hospital. Back then they didn't provide even food or liquids. The doctor himself brought Loren juice and water, but he was unable to drink anything. He lay semi-conscious. Unfortunately, I let the team go back to their homes as soon as the crusade was over, believing that we would be leaving as well. Now I was alone and I would not leave him in this condition, even for a minute, to go out to get food and water for myself. I prayed. The next morning there was a light knock at the door and a young woman came into the room. She said she was a Christian. She had heard about our situation and brought food and water for us. From that point on, for three days, different people whom we had never met, some not even understanding English, came at mealtimes with food and water for me. Loren couldn't eat or drink or do anything to help himself. The doctor explained that malaria thickens the blood and actually creates a situation where you don't desire food or water, although you desperately need it.

By this time, I had found some nurses in the hospital who were Christians. I invited them to the room every night at midnight when no one else was around. We gathered around Loren's bed and softly sang Christian songs and prayed for him. I thank God for having saved me and making me a strong Christian wife who knew how to fight for my husband's life; otherwise, he probably would not have made it through the first night. The doctor ordered a total of five "drips" (IV bags) of quinine during those three days and Loren slowly began to recover.

The Indian doctor released us out of the hospital and rented a room at a tourist hotel for us to stay two nights and three more days, paying for the rooms himself until Loren could get some strength back. We will never forget this wonderful doctor and his kindness to us. After three more days of rest at the hotel, we got into our plane, and although Loren wasn't strong, we flew up the western side of Lake Victoria over Bukoba where we had read in the newspapers that the scourge of AIDS had devastated almost everyone there except young children and old folks. It gave us such a burden for the people and we prayed for the survivors there. From the air, the place looked almost deserted.

We landed in Entebbe, Uganda and taxied past the old control tower. It still bore the bullet holes from when the Israelis raided the airport to rescue their people trapped in the summer of 1976 while Idi Amin was still in power. We remembered the incident and made a point to read the account and find the film "Raid on Entebbe" which tells the story of the counter terrorist rescue mission. We were both interested in history and found it fascinating to be at this place and to see the untouched conditions of that event.

LAKE BARINGO

We stayed briefly in Entebbe; then flew back to Kenya where we met two young men out of one of the churches we ministered in. They volunteered to go out into the bush with us to preach. We were going to the small town of Marigat in Lake Baringo District. One of the main means of making a living in that area was selling honey harvested from killer beehives. It takes a brave person to do this job. The only containers around were old whiskey bottles which the people washed out and poured the honey into to sell on the roadside.

The taste was strong, but good and added something sweet to our diet of local food.

The two young men went ahead of us in an old borrowed pickup truck carrying our supplies to Kampi ya Samaki where we would camp. We had to make a low run over the dirt strip to warn people and animals off so we could land. The lake itself was just to the right of the airstrip and filled with hippos and crocs. We were warned to take care at night because the hippos forage for food after dusk and could wander as far in as our campsite. We parked the plane and set up our camp. That evening we met an N'Jemps, a Masaai sub-tribe warrior named Paul who lived close to the airstrip. He was a strong young man who had a lion's tooth hung around his neck from a lion hunt with other morans, the young warriors of the tribe.

We bonded right away. He was hungry for the things of God and by the next day, we had led him to the Lord. His tough warrior countenance immediately softened into a sweet spirit only knowing Christ can produce. This was to be the start of a long and lasting friendship.

We soon learned the hard way that bad storms blew in every day by late afternoon. Our first evening we bought a chicken and some goat milk from the village and one of the young men with us killed it to fry for our supper. He loved to cook and always asked me to teach him new recipes. These boys seldom had a chance to eat meat and they licked their chops. When the chicken was about half done, one of those bad storms blew in suddenly and we all had to run inside our tents and hunker down. It blew so hard we had to hold the tents up inside with our hands to keep them from collapsing. Water was everywhere inside and out. Everything was soaked, including our clothes and mattresses.

We finally survived the night and the storm blew over. The next morning, our cook came out ready to finish frying the chicken, but I told the fellows it was impossible, because of the heat and delay. We could get food poisoning. It had been soaked in egg and milk and I knew it had gone bad because we were in such a hot climate. I told the guys to throw the whole concoction away. We were all starving and the men began to actually cry and plead over that chicken, but I told them eating spoiled food is why they had stomach problems. I made them bury the chicken. You would have thought a dear family member had passed away they were so sad. They still talk about it to this day.

We spent that morning clearing away the debris around our tents, getting our mattresses out in the sun, and sweeping water out of our tents. I had rigged up a clothesline to hang our wet clothes on. Loren said he would always remember me standing out in front of our tent, the homemade clothesline draped with wet clothes behind me; my hair was up in pink rollers and I was seriously reading *You've Come A Long Way Baby* by Mary Jean Pidgeon. I guess I was trying to figure out where I had missed it.

PULLING THE BANNER

Loren began to set up the banner he would pull behind the plane to advertise our impromptu crusade. The procedure consisted of creating the words from the individual letters and clipping them together to make the sentence on the banner. This was the first time he had ever done this on a primitive airstrip so of course I was a little nervous. We also realized most people couldn't read English or Swahili but it would attract attention. Of course the native grapevine had already been alerted of our presence. The first question in any conversation was "habari gani?" What is the news? We were it.

Part of the set up procedure for the banner, was to put two stakes in the ground, ten feet apart and then set ten-foot long PVC pipes, about two inches in diameter, in each of them. It was designed so the poles would easily come loose after he had snagged the banner and detach from the rigging. Then, he stretched a small cotton rope across the top of the poles. It was tied in a big loop, which stretched across the top of the two poles and was part of a rope stretching out about 150 feet on the ground. The banner was attached to the end of the rope and lay at a forty-five degree angle on the ground parallel to the runway. This would enable the banner to lift off smoothly.

The procedure for setting up the plane to pull the banner is in itself innovative. Loren made a crude grappling hook "holder" out of two small pieces of two by fours. This fit over the pilot's door when the window was opened. It stuck outside the plane's door about nine inches. He wrapped the steel cable around this and stuck the three-pronged grappling hook in a hole in the wood holder. This cable now stretched from his window to the tail where it was connected to a quick release.

Because of the way this holder was designed, Loren had to takeoff with the window open and without the banner. The objective was to snag the rope to the banner from the air. After taking off and gaining altitude, he used his left hand to loosen the grappling hook and un-wrap the cable so the hook and cable would fall and hang behind the plane. Then he closed the window, and circled back toward the airstrip. Snagging the banner had to be done against the wind so after he caught the banner, he would get plenty of lift which would help him gain altitude in a hurry. Loren had to fly the plane down low to snag the rope with the grappling hook. Trying to take off with the banner would have been too much drag.

Believe me, it almost takes the nerve of a Kamikaze pilot to do this, but Loren has never been accused of being faint-hearted and he did not want to die trying it. This trait has sometimes gotten him in trouble. In trying to snag the banner, if he came in too low, he could get the rope in the prop, which would be deadly. On the other hand, if he came in too high, he would miss the rope with the grappling hook. It was a delicate, precise, and white-knuckle maneuver. As he made his approach, all his senses and mine just watching, were at their peak. After flying low over the rope, he pulled up sharply and hit full power, which swung the grappling hook down. He felt a tug on the back of the plane and looking back, saw it lift off the ground. He had gotten it. What a great feeling for all of us. When the banner lifted off the ground, he said he could instantly feel the drag on the plane. He went full power until he got one thousand feet off the ground, then leveled out with it and adjusted his prop and put in ten-degree flaps. This gave him more lift at slower speeds. It looked like everyone came out from everywhere and looked up to see what was happening. We did catch their attention. After pulling the banner around the town three times, he returned to the airstrip, released the banner from the air, then landed with a great sense of accomplishment and airmanship.

AN ATTEMPT

After ten years in Tanzania and having some great crusades, this little meeting was unforgettable for all the wrong reasons. We found a spot in the middle of the small trading center and stood on the back of the borrowed pickup to preach on. We had a generator, a one hundred watt amplifier, and two horn loud-speakers. The two young men with us put the loudspeaker horns up in a thorny tree to the left of the "platform." I opened the meeting with a welcome greeting before singing a song, and then it was Loren's turn for the first message in our first "village crusade". He was the pilot, the advertising agent, the soundman, the generator operator, the technician, and the evangelist.

The first thing that happened was the generator wouldn't start. After much sweat and persistence, it finally started. Loren plugged the amplifier into it and began to preach but soon the P.A. system acted up causing him to have to stop preaching and work on the P.A. awhile. It would start working again and so he would go back to preaching again. There were problems throughout the whole message. Either the generator or the P.A. messed up and the next day, it went from bad to worse. The small crowd stood across the street smiling and appeared to be entertained by our struggle. Finally, after continued problems, he just stopped in the middle of the message and told the guys to "Pack it up; we're going back to camp." We had a lot of issues to work out in this new endeavor to go to the villages.

Chapter 8
Hard Lessons

The magnificent aluminum platform we had built for crusades and to be carried by our DC-3 sat in our garage in the states. It was covered by dirt dauber nests. Loren and I agreed we needed to get it ready by faith and that God would somehow get it to Africa.

I knew when to be still and pray. Loren spent the whole day cleaning it up. That evening after supper we talked about it and having done all we could do, we went to sleep. The next morning around seven a.m. the phone rang; it was a dear couple we had met in one of our partner churches.

They had become good friends and been to the dedication of the DC-3. Brother Jim said, "Loren, I couldn't sleep last night. The Lord talked to me and He told me to send your equipment to Africa."

He didn't have any way in the natural of knowing what we had done the day before.

"Where are you preaching Sunday?" he asked.

It was in a town not far from where they lived. He said, "We'll meet you there and bring you a check."

When Loren told me what had been said, we both broke in emotion and praised the Lord. We were convinced if he had not prepared the platform in faith the day before, the Lord wouldn't have spoken to that dear brother to send it to Africa. Truly, both he and Loren were led by the Lord to do what they did.

Now this miracle of getting the funds to send the equipment to Africa brought us another problem. We didn't have a truck to carry the platform, P.A., and other equipment, once it all arrived. This was no minor issue. Sometimes you need a bigger miracle to keep the miracle you get.

We preached that morning a few hours away from our home. This church and the Pastors had also been such great friends to us and had supported us for many years. As planned, the man and his wife who called met us at church and afterward we all went to lunch together. We never mentioned the need for a truck. As we walked out of the restaurant after a good time of fellowship, this dear brother and his wife handed us a check to send our equipment. Then with tears in his eyes he said, "The Lord talked to me again last night. He told me to buy you a truck too." You could have knocked us down. Only the Lord knew of our conversation and concern about the truck. We all greatly rejoiced in the Lord. Before long we had gotten a container and our equipment was on its way to Africa. "Without faith it is impossible to please God". Hebrews 11:6.

DEEP TREACHERY

A year had passed since we brought our Cessna 206 into Kenya. The government had been kind enough to allow us to keep our plane under U.S. registration and we had it serviced regularly at Wilson Airport in Nairobi for annuals and any other issues.

Now we were able to buy a four wheel drive, seven ton Fiat Iveco Truck outfitted for the bush thru a local pastor. It previously had been used as a mobile medical unit and was exactly what we needed. This would set us free to move about with our crusade equipment and evangelize the villages on a much larger scale. But soon after settling into Kenya, we came across the same kind of corruption in the church as we had seen in

Tanzania. Too late, we discovered the truck had belonged to the government and didn't have a real logbook (title). The papers that were given to us were virtually useless. This meant the pastor and the elders had connived and taken thousands of dollars from us and left us with nothing.

Despite men sabotaging us, God worked with us, however, we could hardly sleep at night. These people had set us up to take what money and equipment we had. It was an excruciating spiritual battle while preaching a small crusade in town, fighting not to lose our truck, and then dealing with customs trying to get our container out; all at the same time. It was a nightmare. After all we had been through in Tanzania, having recently survived a terrible plane crash and the loss of our big plane; it was like the vultures had swooped in to finish off what the lions and hyenas hadn't eaten.

Loren went to Nairobi struggling to get our container out of customs with no results. When he returned, in desperation, we visited a local missionary we had met on an earlier trip and shared our dilemma with him. He called a friend of his who was influential in the government for advice. We later found out this treacherous pastor who had made the bogus deal with us on the truck, had also schemed with the elders of the church to try to have us deported. If he succeeded, he might have taken possession of our airplane, the aluminum platform, our big P.A. system, the truck, and the funds we had given for it. Now, this influential person intervened for us with some top government officials and this put fear in the hearts of our adversaries.

Later on we were told that a few years earlier this pastor had already taken a P.A. system from another missionary, so what he tried to do to us was not new territory for him. This was his true character. Satan was after us and he had plenty of little helpers. "And no marvel; for Satan himself is transformed into an angel of light. Therefore, it is no great thing if his ministers also be transformed as the ministers of righteousness; whose end shall be according to their works." 2 Corinthians 11:15.

The ministry had been stymied for six weeks dealing with these serious issues. This situation was eating up our five-month tour. I would teach in ladies' meetings during this time and in "lunch hour" prayer meetings in town. It was so outrageous what was happening to us, we almost thought it was an illusion. Not only was the threat of losing everything such a nightmare, but also it would be a huge blow to our partners who had paid to get the equipment shipped to Africa and provided the funds to buy the truck. There was a lot to fight for. God gave us dreams at night and showed us exactly what our enemies were planning next. As we discussed these dreams in the morning at breakfast, the Lord would confirm them to us by His Spirit thru other people.

AN IMPROMPTU MEETING

On a Friday evening during this time, Loren told our small team, "I'm tired of sitting with all this mess going on wrangling over paperwork and not being able to preach." We were limited in taking the truck out because of the logbook issue. The local government had approved us staying close to town only while they tried to help us resolve this issue. "Let's start doing what we came here to do. Let's take the truck and go preach somewhere nearby tomorrow."

Our small team brought out a map and Loren randomly pointed to a nearby town within the confines of our driving permit. We were told it would be about a two-hour drive. We agreed, "Let's do it." We knew absolutely nothing about the place so the next morning we sent one of our helpers out early to try to connect us with a pastor who was interested in winning souls. He ended up finding a small church and talked to the Pastor about our coming to preach there that afternoon. The Pastor told him that last night his church had an all-night prayer meeting and marched all over the town praying and claiming their town for Jesus. He took our coming as an answer to their prayer.

THE ANGEL

We had designed and constructed a portable platform for the right side of our truck. It folded up and locked at the top of the truck's van like box. When we needed it, it folded down quickly and we could preach on short notice. Soon after we arrived, we set up and began singing. A few hundred people came and scattered about, at a distance. The Pastor interpreted for us and a fairly good number accepted the Lord and some wonderful miracles happened. A young woman came to the platform and testified, saying she knew we were on our way. She said, *"Three months ago* my angel told me Loren Davis was coming."

We were not known in Kenya at this time.

She continued, "He even showed me your truck." She said, "He told me to bring my six year old daughter who was crippled and had never walked and God would heal her. He told me the day you would be here, so I came from my village last night. When I saw your truck driving in this morning, my angel said, 'This is Davis's truck. He has come.'"

Incredibly, she was told this before we had even bought the truck and had no ministry markings painted on it yet because of the legal controversy. This encouraged all of us and we wondered. Her daughter was healed that afternoon and walked for the first time in her life. She began to walk around town with a big crowd following her singing and praising God.

THE SABOTAGE CONTINUES

It was on our hearts to preach in unreached villages in the bush because by this time we realized virtually no one went to the remote areas to take the Gospel. We wanted to go. Our enemies wouldn't give up and when this enemy pastor heard we were making these short trips, he tried to stop us from even going out to the bush to preach. He continued to try to sabotage our ministry but didn't know that he was up against a couple of missionaries who would not lie down and give up. "The righteous are bold as a lion". Proverbs 28:1.

We had great confidence because we knew God was with us. This man went to the Provisional Commissioner, who is like a governor of the province, and told him we were bad people and should not be allowed to go into the bush to preach. A lot of people think you can come to Africa and start preaching and doing whatever you want, but there are many legal technicalities you must tend to first or you can get into serious trouble. It can be complicated. Fortunately, our work permits were in perfect order. The denomination we were working under was also helping us in this battle.

This pastor's interference forced Loren to go see the Provisional Commissioner himself. After visiting with him about who we were and what we wanted to do in Kenya, we gained his favor and he released us to go to the interior to preach. The Bible is so true: "with every temptation he also makes a way of escape, that ye may be able to bear it". 2 Corinthians 10:5.

Through a great struggle and a lot of perseverance, God miraculously helped us to finally gain full legal possession of the truck and also to get our container filled with our crusade equipment released from customs. There were no words to express our relief and joy. The scheme of this pastor and his elders failed.

We were invited to return to that little town we pointed to on the map for a small crusade. We accepted the invitation and brought with us some of our new big speakers. A crippled young woman was healed by the power of God and walked home in her own strength without her crutches. When she got home, her parents were amazed. Everyone knew this was the hand of God and many believed and were saved. There was a man who had a deaf ear from birth and instantly received his hearing. A mama brought a small boy. She told us he was born with a white film over his eyes and was blind. The white film had disappeared after prayer and he could see. A young woman, who could hardly walk, now walked; a man and woman nearly blind could see; deaf ears were opened. Almost everything happened.

We can't heal anyone. Only Jesus, the Messiah, can do such things. We are only workers together with the Lord. He confirms His Word with signs following. "Go into all the world, and preach the gospel to every creature. ...and these signs shall follow them that believe..." Mark 16:15-18.

Many attended and we were told the crusade could be heard through our big speakers for miles away on the hillsides. A drunk who couldn't even see the crusade because he was so far away, came out of a bar hearing the sound. He repented and turned to the Lord; then came to the crusade and testified. He said he went out and bought a bottle of booze and showed it to a friend and then poured it out onto the ground to prove to him that the gospel had changed him. He said this was the last bottle he would ever touch. The Pastor's church was packed out with new converts. He said people stopped him on the streets wanting to get saved.

KERICHO

We continued working and took the ministry to another small town in the nearby mountains to the town of Kericho. It is a beautiful and lush area with tea plantations everywhere. A precious older African mama asked us to stay with her at her house which had been an old colonial home. She was a lovely person and we were intrigued with the tribal markings cut on her face. They had obviously been carved there when she was a young girl and we marveled that God could reach down into a pagan society in his mercy and save. The roads around Kericho were unbelievably muddy because of the incessant rains and we had to leave our vehicle parked either at her home or if we were able to get out, would invariably have to walk back in on little foot trails through the jungle, plugging along much of the way through mud and wading through water for several kilometers in the pitch dark. After the rain, a tractor was able to pull our vehicle out, but by the time the services were over, the rains would come again and we would have to walk back through the mud again to get back to the house. This is where we were first introduced to "gum boots" which were sold in that area. The pastor and Loren talked about the mud and rain, but I began to pray, "Tonight, our road will be dry." Astonishingly, when we got to the turn-off to go down to the house, our road was dry. There was rain and mud everywhere else, but not on our road that night. It was dry. We learned that Jesus said, "We have not because we ask not," James 4:23 is true.

THE SISAL PLANTATIONS

We began to penetrate deeper into the villages working with local pastors in the sisal plantations. These were located in an arid area and were owned by Greeks. Thousands of acres were cultivated by workers who lived on the properties and made Rope out of the plant fiber; a very interesting process to watch. The workers on these plantations were provided housing but were extremely poor. We soon fell in love with these precious ones who lived so far off the beaten path. It was a privilege to preach the saving Gospel to them and see so many come to Christ. God continued to confirm His Word with many miracles. A crippled lady was healed; an old man who was carried by three men to the meeting was healed and danced all around; a lady who stood in prayer for her son who was paralyzed later testified, even though he wasn't at the meeting, when she went home, he was healed the same day. He was supposed to be admitted to the hospital to have his legs amputated. The surgery was called off. What a testimony. The word of God is so powerful. Our job in preaching Jesus Christ is to build faith for believing who He is, He is God, and that healing are signs of His power, His ability to deliver us from the fear of death and to give us hope for a future with Him in Heaven. "Jesus Christ, the same, yesterday, and today, and forever". Hebrews 13:8.

AN UNKNOWN VILLAGE

We were told we were the first white people to ever come to this particular village. A local Pastor brought us in and introduced us. The people just stared at us, wondering. We told them the God who made the heavens and the earth sent us and we preached about His son Jesus, whom God sent down to earth because He loved them so much. He could take away their sin and put His law in their heart. Many accepted Christ and miracles happened as we read from the Bible and the interpreter followed in the local language. A child born dumb began to speak. An insane woman came back to her right mind. That is why Jesus said "Believe me that I am in the Father, and the Father in me: or else believe me for the very work's sake." John 14:11. God made Himself real to them because they believed on Him.

ANAMOI VILLAGE

In the village of Anamoi, a witchdoctor got saved. He testified he had carried around a human head in a bag and would even drink human blood in his worship to the devil. After hearing the Gospel, he gave up witchcraft and all these practices to serve the Lord. He was also baptized in the Holy Spirit. Here is someone who understood the power of the dark side of the spirit world but came face to face with light, a greater power, and was changed. Jesus said, "I am the light of the world: he that followeth me shall not walk in darkness, but shall have the light of life". John 8:12.

DAR ES SALAAM

We were invited back for a major crusade planned in Tanzania. This time there was strong backing behind us in the capital city of Dar es Salaam. The venue agreed on by the crusade committee was a field directly across the street from a Muslim mosque. Our posters and handbills were distributed and ads were placed in the newspaper and on radio. Our team, truck, and equipment were brought from Kenya at great expense. The new aluminum platform was set up for the first time and we excitedly prepared the field for the meeting, however, the day we set up the platform, a terrible blow hit the country. The father of the nation of Tanzania, the first President, died. This immediately threw the whole nation into mourning.

The government of Tanzania invoked a moratorium on doing business and from conducting any public meetings to be enforced for one month for a time of national mourning. This instantly halted our crusade efforts for at least a month and put us in a terrible financial predicament. All of our money on advertising was now useless and we had to support our big team for an extra month while waiting out the mourning period. We thought of returning to Kenya but one of the major bishops who had invited us advised us if we left Dar es Salaam that by the next day, everyone in Africa will have heard, "Davis came and didn't have the strength to stay." It was a terrible predicament.

Another friend of ours had flown over from the states to support the meeting, but arrived sick. We were concerned about his health, but he insisted on staying and would not see a doctor. Now with the city at a standstill, we looked for a place to stay for a month or longer. Our finances were tight and we had a big team of men to house and feed as well.

Dar es Salaam is a port city on the Indian Ocean and had a very large old landmark called The Kilimanjaro Hotel. It faced the waterfront and you could watch the ships from different countries coming in and going out of port. It was intriguing to us from the West because many of the ships were from nations like Russia, the eastern bloc nations, as well as the North African Muslim countries. We had always admired the hotel because it was so stately. It had visibly deteriorated over the years since the colonists had built it a hundred plus years ago, but it still carried the mystique of old Africa.

One day, we saw a tiny sign of life around the hotel and decided to investigate. The hotel was almost empty but had staff at the front desk. We asked if they still took guests and possibly had a large suite available. They said yes and made us a great deal to stay for a month. It was a two-bedroom suite with a balcony overlooking the waterfront. There was a large living area between the bedrooms and three bathrooms. We had two single visitors with us and didn't like to have them separated from us in a strange country, so it was a good fit if we pooled our resources together. Everyone agreed and we took it. Our female visitor took one bedroom, we took another, and the single gentleman had the living room couch with his own bathroom. We felt so thrilled to stay in this old place and at such a bargain price. We were able to have our Tanzanian pastor friends come to visit us in groups with plenty of room and have our team meetings in privacy. It was also a great place to pray on the balcony overlooking the harbor.

This living arrangement turned out not to be as good a deal as we thought. The first thing we had to come to grips with was the elevators didn't work and we were on the fourth floor.

The restaurant was also not functioning but that was not our biggest problem. Getting water was sporadic and when we got it, the water was rusty because the pipes were so old. We soon learned the hotel was in receivership. The government now owned it and was looking for a buyer. It had become a white elephant. The few staff that remained rented out a few rooms to try to keep a job for themselves and that's where we came into the picture. We finally found a staff member who agreed to make us tea and coffee, but soon learned the water was not safe to drink and you can imagine what that did to us. With all its problems, there were blessings. We had paid our one-month lease up front and because of that, the staff was able to buy breakfast food for the few of us staying there. They told us what a blessing it was to have us because it meant they had a job and a little money to feed their own families.

During this time the Muslims in the mosque across from where we were set up for the crusade became aroused against us. They didn't want any crusade to be held there. One day, about three hundred men came and threatened to burn up our equipment. We had hired men to be watchman for this interim period and they were attacked. Our men fought back. When it was all said and done, our men had whipped their attackers and the others fled the field. It was a volatile situation. We decided it was best to petition a change of venue for the crusade because of the continuous threat of violence. Thirty-two days after the President's death, the official days of mourning were over and the city officials gave us permission to hold the crusade at another location, but now, they only agreed to give us two days instead of the seven we were initially promised.

Loren pulled a banner behind the plane over that great city of Dar es Salaam to draw attention to the crusade and put a few ads in some smaller newspapers and added some radio ads. Despite only having those two days to preach, God showed Himself strong. Many people came and gave their hearts to the Lord. A man testified God healed him of elephantitus and the massive swelling had completely gone down in his body. A ninety year old man accepted Christ; a girl who was paralyzed on her left side was healed; and a man with a blind eye received his sight.

On our walk to the hotel one day we met a man from Dodoma who used to be a Muslim. He told us he had been saved in our crusade years ago and was excited to see us again. We asked about his family, but he said he was forced to move away because they threatened to kill him for becoming a Christian. Despite the sorrow of his family rejecting him, he said he would never turn his back on or leave Christ. He wanted us to know he was still saved and following the Lord. Another pastor in Dar es Salaam testified a man from his church told him he was at our crusade in Kigoma-Ujiji. He said his daughter who had been deaf and dumb had been totally healed. We don't always get to hear everything that the Lord has done, so these testimonies were wonderful to us. We are just the mouthpiece. He is the Saviour and the Healer.

Our truck driver at the time was a man from Singida, which is in the north central part of Tanzania. He and our soundman were away from their families traveling and ministering with us for five months. Many in our team suffered while serving the Lord with us in this tough field but never complained. They loved being a part of what God was doing.

At the end of the crusade in Dar es Salaam, we held a pastors and leaders conference as part of our program. The topic again was the issue of Sound Bible Doctrine. Titus 2:1 exhorts us to "Speak thou the things which become sound doctrine". 2 Timothy 4:3 tells us, "For the time will come when they will not endure sound doctrine; but after their own lusts shall they heap to themselves teachers, having itching ears; and they shall turn away their ears from the truth, and shall be turned unto fables." We had a very great turnout, but the conference sparked a lot of heat because we dealt with the corruption within Christianity. During this conference, a Nigerian pastor came to the microphone and argued that he was "equal with Christ"; that he was "a god." The continuation of that scripture is "but ye shall die like men, and fall like one of the princes." Psalm 82:3. Although this doctrine is totally opposite of what the Bible teaches, it was a great illustration to us on how far some Christians had fallen in their understanding of scripture.

A few pastors were very concerned about the many false teachings in Christianity and appreciated the conference, but the majority rebelled and rose up against us. Many corrupt teachings had arisen such as: a perversion of the plan of salvation, and the abuse of the prosperity message. The "Sow your seed" gospel was very big and growing. Money was virtually the *only* thing most pastors taught so the parishioners were kept in bondage. These perversions had come to Africa from the many popular preachers from America and Europe. Lying, dishonesty, stealing, and false doctrines had seemingly taken over most of the church. A spirit of greed was perverting the true Gospel. It was in this meeting that we realized Muslims were not the church's biggest enemy; it was corrupt preachers and corrupt teachings within Christianity itself. These pastors joined together and ran a full-page article in a newspaper against us because of our Biblical call back to sound doctrine. Their money train was threatened and they, in turn, threatened to destroy our ministry if we ever returned to Dar es Salaam. These experiences are what led us to write the book, *The Paganization of Christianity.*

We began to realize how global these perverse teachings had gone. An Indian evangelist came to Kenya not long after that conference in Dar es Salaam for an "interfaith" crusade. It made the front page of the newspapers. This Indian evangelist openly called for Hindu, Muslim, animists, as well as every brand of Christianity to join together in collective worship. This perverse ecumenism had stretched Christianity to sheer absurdity, literally putting all gods on the same level. This was done in the name of "peace". We began to see the *global* trend to clearly de-throne The Lord Jesus Christ to the level of other so called "gods". Our hearts became very alarmed. Exodus 20:1-2 says: "And GOD spake all these words, saying, I am the Lord thy GOD...thou shalt have no other gods before me....vs 5 for I the Lord thy GOD am a jealous God". He makes this statement six different times in the scriptures.

Precious little one finds joy in a flower blooming
in the dry land.

Elephants and Lions in the world famous Maasai Mara
National Reserve of Kenya.

Pokot man. Note the painted clay yarmulke on his head.

God's handiwork on this Zebra shows we are all unique and one of a kind.

Maasai man and woman.

Celeste with Samburu Women.

Giving Swahili Bibles to the East African community.

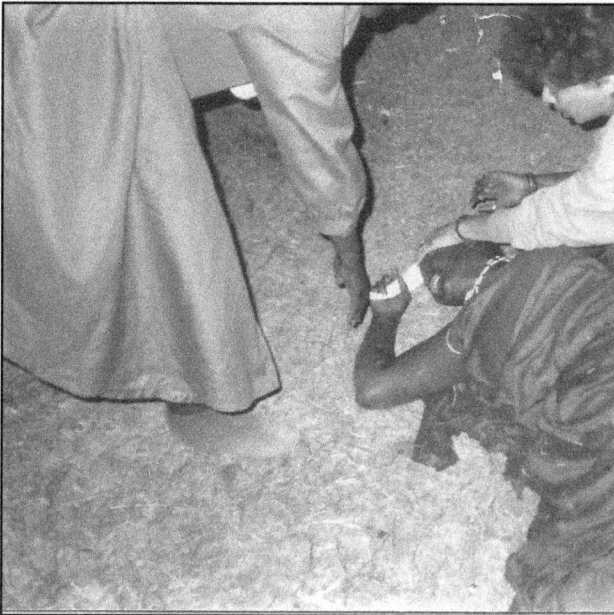

Casting out demons that have manifested during a crusade.

This Congo Crusade drew an estimated 400,000 people.
Loren invented the lightweight portable aluminum platform.
It breaks down into 8ft sections and can be easily packed into
our trucks and ready for the next meeting.

Hands are lifted in surrender to the Lord while thanking Him
in faith for the salvation of their eternal souls.

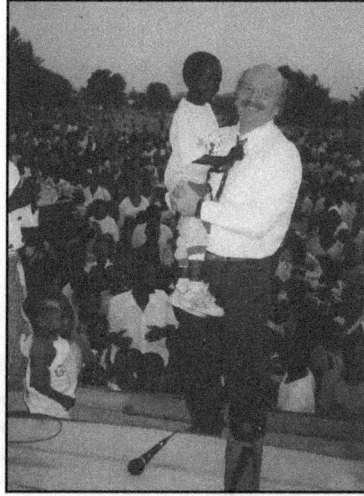

This woman was healed and lifts her crutches above her to show what the Lord has done. He is the same yesterday, today and forever.

Jesus said to have the faith of a little child and to let the little children come to him. They are among the first to run to the altar in our meetings.

Raising your hands is just agreeing with God that we need a savior. These precious people are making a conscious decision to follow the Lord Jesus Christ.

Singing in worship to God and welcoming the people to hear the Word of God.

CNN interviews Loren just before he preaches at the Kibera Crusade in Nairobi, Kenya.

The Dandora slums crusade in Nairobi, Kenya.

The early years in Tanzania crusades.

Feeding the hungry. I just love working for Jesus.

Thankful Pokot women. Our truck in the background is filled
with grain. We give food for the mind, soul and the body.

Jesus loves the little children of the world.

We had fresh Impala for dinner that night, roasted in an open pit dug in the ground.

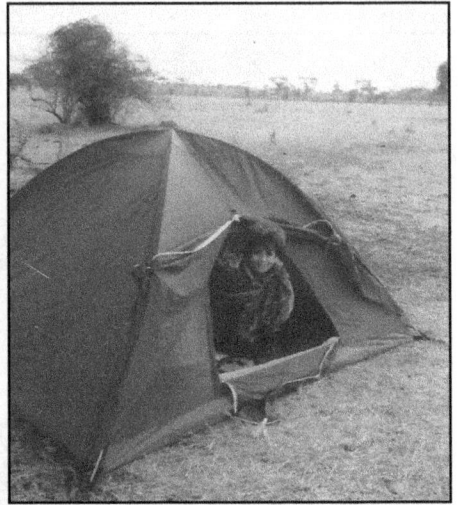

That's me peeking out of my
home in the deep interior
of Tanzania.

My sweet little friend is waving
to you and saying "Thank you"
for sending me to tell her about
the savior of the world.

Loren is so full of joy sharing the Gospel in
the villages.

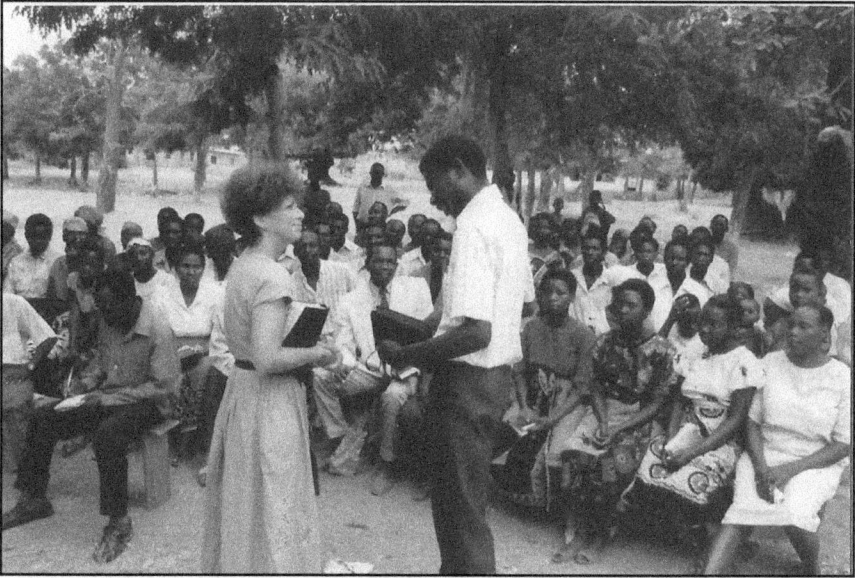

Having church out in the open under a tree.

The joy of giving a place to worship with protection from the sun and rain! These churches also serve as pre-schools. for the village. Thank you partners!

We found ourselves caught in the war between Hutu and Tutsi in 1994. We ministered to the refugees in Rwanda and the Congo.

This was an amazing trek into the land of 1000 hills of
Rwanda to track the mountain gorilla.

Chapter 9
The Villages

In Loren's mind, he was exclusively an evangelist and I, a teacher, and we did not want to entangle ourselves with anything else. However, in our early travels in Tanzania, we noticed whether going by car, train, bus, or plane, there were innumerable villages in the interior that were unreached. It was the same way in Kenya and the other East African countries. We had often commented to each other when we saw this lack and I think God was working on us way back then to see the need and he was developing a vision in us to fill it. God didn't want to leave these interior villages to perish without Christ; that much we knew. "For God so loved the world…" John 3:16-17.

It was then that this vision for actually building churches began to get a hold of us. We had seen some western styled churches begun by enthusiastic people, but ten years later, they were still unfinished for lack of funds or because the funds had been pilfered off and pocketed. The idea we had was to build many simple and inexpensive churches. This way, many villages could be reached and have a "Village style" church building to meet in and the Gospel would spread quickly. We were in total agreement on this plan.

As we began to mention this vision to friends in the states, funds began to come in from different churches and individual partners to sponsor the village work. This was the beginning of something we were later told by Africans was unprecedented and wonderful. Our experience had also shown us that there were many men coming out of Bible School with no funds to start a new work. We realized that matching a pastor with his tribal mother tongue and village upbringing would be a way to help each of us to fulfil the call of God on our lives. The denomination we were working under supported the idea and it would expand to other bible believing denominations in the country.

THE POKOT TRIBE

We built our first churches among the Pokot Tribe in a place called N'gingyang, upcountry from the Lake Baringo area in Kenya. The Pokot were a pastoral people but had a violent reputation as we had mentioned earlier. They had been unfriendly on our first visit a year prior because they had plenty of grassland for their herds and were fat and happy so to speak. They did not feel a need to hear from foreigners about anything. The Pokot didn't consider their land as even a part of Kenya and were virtually ungovernable by the Kenya authorities. The people were wild and unpredictable if you did not understand their ways. When one of their elderly was unable to function and take care of himself, the custom was to take them out to the bush and leave them to the hyenas. If the hyenas didn't eat them, it was viewed as a bad omen. However, that year, we found them in the midst of a terrible famine. Many had starved to death. It had not rained for almost two years and the carcasses and bones of goats and cows were everywhere. The riverbed was totally dry, but we saw the women and children digging in it, trying to find a little water deeper down in the dry soil. Now they were doing anything to survive. It was heart wrenching.

We came in with our truck packed with maize to give to the starving people. This visit they were open to us helping them. It was an amazing pattern that evolved; in every village we came to there would be a big rain right after we preached and gave out food. Rain was considered the favor of God and they began to associate our coming with the gospel to the drought breaking in the villages. This happened in village after village, day after day. The people took this as a sign that God was with us because we had brought the blessing of rain. Then, as we preached the message of God's love for mankind in sending His Son to save them, we prayed for them and it appeared whole villages, one after another, responded to accept Christ. The Lord did many mighty miracles among them. Many who were blind testified of receiving their sight and Loren would ask them to follow his finger, moving it indifferent directions to check what they were seeing. This was the beginning of a phenomenal breakthrough among these ferocious people.

Since we were deep in the interior, we had to pack-in our own supplies including water, food, chickens, and tents. We had specially made water tanks attached to our truck and also carried in extra barrels of water and fuel. There was no good place to pitch our tents, so we ended up in a stony open area, which we soon discovered was full of scorpions. These scorpions were opaque in color and blended perfectly with the stones and white rocks. It was a stiflingly hot area so we wore open sandals, which made our feet more vulnerable to the scorpions. We were under a completely open sky. In the daytime the hot sun beat down on us. It was so blistering hot in the middle of the day, the Pokot came and mercifully showed us an old dried up river bed close by with a few small bushes around it. They motioned for us to dig in the ground until we hit moisture and lay there most of the afternoon trying to cool off a little to survive the day. It was unmerciful heat and we had a good taste of how hard their life was. It was only the grace of God that enabled us to endure it. Because of the heat, we would go to the villages very early in the mornings and then late in the afternoons to preach and distribute food. Mid-day was unbearable. We suffered until sundown and then only late at night would we find a little relief.

We found out quickly indeed we had been called by God to go to this place; otherwise, we could have never endured it.

I was very intrigued with the people and noticed interesting things about the Pokot. I enjoyed observing them. For example, one morning we brushed our teeth in the usual way with our western toothbrushes and toothpaste, but I noticed the Pokot had a small green branch in their mouth. After investigating, I found out they used a short branch from the Sogotiewe tree. We are using the word they used for the tree and the spelling is by sound. They would break a small branch off the tree the length of a toothbrush and chew on the end until it made a brush. With this, they brushed their teeth. This special tree created an enzyme that made their teeth beautifully bright and clean. We soon used that method ourselves and found it was better than any high-priced tooth-whitening system the West had to offer. It also made the gums tight and healthy. Unfortunately, the tree was only found in certain areas. I had to go to the dentist later that year for an annual checkup while we were home in the states. The hygienist was astonished not only at the whiteness of my teeth, but the condition of my gums. Previously, I had a problem with periodontal disease, but now the condition had greatly improved. They all wanted a branch of that tree for themselves.

Another bush created soap-like qualities from its leaves when mixed with a small amount of water. Even if you did not have water available, rubbing the leaves on your skin formed enough soapy moisture to cleanse it. It was amazing how the Lord created the trees and bushes. We thought of the projects in the west to send shoeboxes with toothpaste, soap, and other toiletries to these "poor" people. Though well intentioned, those boxes were needed in the city slums. God had already solved these issues for these bush people within His magnificent creation. "And God said, I have given you every herb bearing seed which is upon the face of the earth, and every tree, in which is the fruit of a tree yielding seed; to you it shall be for meat." Genesis 1:29.

THE FAMINE

We encountered a precious old woman who was skin and bones. She stooped over and picked up a few grains of maize that had been dropped from our maize bags while our men off-loaded the truck. Tears came to our eyes as we watched her painstakingly gather every single grain she could find. Believe me; we blessed her with an extra generous portion. Hundreds of people lined up after our services everyday as we gave out the maize to them. What a privilege to help these hopeless people with physical food as well as with the Bread of Life, the Word of God. "And Jesus said unto them, I am the bread of life: he that cometh to me shall never hunger;..." John 6:35.

One afternoon while we tried to cool down in the riverbed, three Pokot women joined us. Pokot at that time were still one of the few tribes that continue to wear their traditional dress of cowhide skirts and beads. The women were topless or covered with a skin but loaded down with beads like a great collar around their necks so you hardly noticed their bodies. Many people may not realize that not wearing clothes is not a preference, but a fact as they live in the wild and do not have access to any other clothing. Their cows provide everything. One woman came to camp one day asking for a dress. I gave her one in exchange for the piece of cowhide she had hung around her neck. On this particular day, these women came and sat in the riverbed with us and through our friend, mama Chipkor, who interpreted, poured out their hearts to me and told how their husbands beat and abused them. Mama Chipkor was the wife of our friend Brother Rugut who had helped us get our papers on the truck. They were Nandi people and working with the Pokot as home missionaries thru the African Inland Church (AIC) in Kenya. They knew the Pokot language and we were so blessed that God brought them into our lives for so many reasons. Yes, as interpreters but they became dear friends. They loved Jesus and we were kindred spirits and knit in the Lord.

Pokot women lived in polygamous marriages and I was able to minister to them and that was the beginning of a long friendship with these precious ones who came to name me *Cheminingo* which means "woman in a small body". They attach your first born child's name so you are never called by your given name. If you are a grandmother you are called *Koko which means* "grandmother" and they attach your first born grandchild's name. These designations made clear to everyone who you were and who your family was.

This was the beginning of many Pokot children born over the years during our stay whom they named Celeste or Davis. This was also their way of keeping records since they don't have a written history. They keep generational records orally by naming their children after some special event that happened around the time of their birth. In our case, it was because we were there. Some were called "morning," "afternoon," or "evening," in Pokot of course, according to the time of day of their birth. It was an honor to have so many Celeste's and Davis's, if it was a boy, running around. This showed they accepted and cared for us.

We had developed a relationship with our dear friend and Bishop of the ministry which we were registered under in Kenya. We built the churches under his denomination and covering. The new churches had to come under a legally registered Kenyan denomination. It was also a way to have the pastors be accountable to the denomination for the work going on in the villages. By this time, all our government paperwork had been completely approved.

GIDEON'S DAUGHTER

Pastor Gideon was the pastor of a broken down, termite ridden AIC church. He was tall and lanky, at least six foot four, and so boney he looked almost like a human skeleton. Gideon was in desperate condition.

One evening Pastor Gideon came rushing over to our tents to see us with his little daughter in his arms. The famine had taken many lives among the Pokot. He told us his four-year-old daughter had malaria and a high fever. She had not eaten or drunk anything in a couple of days. Immediately I went to our meager supply of drugs and brought out and crushed a half tablet of Fansidar mixing it with water in a teaspoon. I tried to gently get the girl to swallow it. She was so frail from mal-nourishment and was undersized for her age. We prayed as I tried to get her to swallow the medicine, but she couldn't keep it down. Concerned, we jumped into Rugut's old pickup and went to look for help. Out in the bush, it was sometimes pos-sible to find a nurse stationed in one of the government dispen-saries, but they were few and far between and usually had very few drugs, if any, available. This baby needed a "quinine drip," an IV with meds for her malaria. It was a dark night. When we finally found someone, it was to no avail because they indeed had no medicine. Now, it was either we believe in God for a miracle or the little girl would die. We prayed and rebuked the spirit of infirmity of malaria in Jesus' name and gave the child back to the father. The next morning, Gideon came to our camp rejoicing and told us the fever had broken that night and his daughter drank and ate a little. In time, she completely recov-ered. Thank God for Jesus, our healer, deliverer, and friend. From that point on I was determined to try to get medicine donated for us to carry to the bush for emergencies like this one. Whenever we would go into town for supplies I would make the rounds of the chemist shops and ask for donations of malaria drugs and simple treatments that could make a difference along with prayer in the life of someone in our path.

POKOT CUSTOMS

The Pokot are a unique people. The women and girls of the tribe wear a necklace called a *shanga*. The size and color of it signifies their place in the society. Mothers hope that their babies and little girls will have at least one string of brown beads around her neck, although we saw many girls with nothing; they were so poor. This custom of beads also signifies the wealth of the father. The father provides these beads to his daughters to beautify them. The more beautifully the girl is decorated, the higher dowries of cows, goats, and camels he can ask for the bride price.

As she gets older, the *shanga* becomes larger. It is always a work in progress; culminating in a large brown *shanga* by the time she reaches puberty. These big circular necklaces can be up to twelve inches or more in radius around the neck and can extend well beyond their shoulders. This necklace is made from a certain tree branch, which is cut into beads with a hole borne through them. The beads are then strung together with thin strings from the sisal plant. At intervals, the beads are held in place by a piece of bone until they get the large effect to stand out. We saw young girls with their mothers helping them cut limbs to begin making their beads. In the process they would cover the string of beads with red mud if available to keep the wood soft and supple. Obviously it took a long time to make. These necklaces are beautiful and mean the girl will soon be eligible for marriage.

The Pokot circumcise both boys and girls. We will tell you about the boys a little later on. Girls going through circumcision are those who have experienced one menses. The circumcision ceremony publicly declares them to be a "woman" and now eligible for marriage. Women wearing feathers in their headpiece indicates they have daughters who went through the circumcision ceremony. This is a sign of honor to the mother. Female circumcision is genital mutilation, a practice done by many African tribes. This ritual is actually outlawed by the government, but they have no control over the people in the interior who continue to practice this custom.

Married women wear colorful *shangas*. These colored beads are purchased during market days when the villages get together and sell and trade their cows, goats, camels, and honey with the traders who come through. This is also a time of fellowship. It can be a weekly or a monthly market and is usually festive and the community comes from far to participate and catch up on the latest news. It was one of our favorite times to mingle with the large groups without calling a special meeting.

PREACHING TO PRIMITIVE PEOPLE

Loren had just finished preaching to the Pokot from the small platform on our truck, when a woman screamed hysterically and ran away. There was a great commotion and the woman's five-year-old daughter came to the front of the crowd and brought me a limp baby girl. People around us thought the baby had died. Loren had just preached to them about the saving and healing power of Jesus Christ. We took the baby and rebuked the spirit of death in the name of Jesus and then I took her inside the truck, while Loren continued ministering to the people. The young sister followed me. I held the baby close while praying for her in the shade of the vehicle; I dipped my finger in water and gently pried open the baby's mouth trying to get some fluid through her tightly clenched teeth. Soon, the child moved and I peeled an orange and squeezed the liquid into the child's mouth. It was a tremendous testimony to the Pokot. After a time we gave the baby back to her big sister and off she went into the crowd to find the mother.

We had a limited amount of maize in relation to the amount of people there, but tried to give everyone at least a bag full and sometimes more. There were hundreds of starving people. They would line up with their cloth bags in their hands, a literal human chain of skin and bones. They were grateful and it blessed us to be able to help them. We had to be watchful because many would cheat and get back in line for a second helping before we had given to everyone. Many of them had accepted the Lord, but they were not all sanctified yet.

This was the beginning of numerous trips to the Pokot, ministering to them, feeding them, and building churches. They made Loren an elder of the tribe in a special ceremony and gave him a Pokot elder's stool to symbolize this honor. I was made a mother of the tribe and given a specially made "mother's belt" to seal my position. We were told if the Pokot went to war with another tribe, that "mother's belt" placed down on the ground could stop a war. We were honored to be a part of their tribe. It meant we as people, as well as our message about Christ, had been respected and for many, accepted.

A FLASH FLOOD

We had camped under some of the few trees by the old dried up riverbed near the crumbling Pokot School. We could tell by the terrain that this area had seen many flash floods in the past. The riverbed was ragged where the earth had been torn savagely, making cliffs along the riverbed up to twelve feet high. In the middle of the night, I heard a mighty roaring sound of water and quickly woke Loren up. We were concerned the river might flood over the bank and cover us or rushing trees and debris could knock our camp and us away. The river rose due to heavy rains coming from the mountains upcountry. It was a dangerous situation and pitch black and we only had a flashlight. There was little we could do but pray. Sure enough, the next morning we found a powerful flash flood had rushed by us in the night. Thank God the river did not rise high enough to reach our tents. The flash flood was so powerful it violently carried with it many large trees and boulders as if they were nothing. The waters had eroded the land all the way up to the small Pokot School and it sat precariously on the edge of the cliff. We used our truck and team to help the Pokot to haul many loads of rock. We weren't engineers but they hoped that by piling them at the base it might reinforce the cliff trying save the school when other flash floods came.

THE NIGHT OF THE SPIDERS

One evening while visiting around our campsite the ground suddenly came alive. It was literally covered with spiders. The men wanted to step on them but first of all they were so many and then the Pokot warned us not to do that because they said these spiders would turn and attack. By this time, I had jumped up on top of our makeshift table to get away from them. It was almost sundown and we decided to make a mad dash for our tent. We unzipped it quickly and ran inside; searching to make sure no spiders had gone in with us. One thing I had told Loren in the past was my requirement for staying in a tent; it *had to have* a floor and could be completely zipped up. He also thanked God for those rules. We don't know if the barometric pressure changed or a small earthquake took place; we only know the ground shook and opened up with hundreds of large white spiders pouring out.

For much of the night, we could hear the spiders scurrying all over the tent. The next morning, thank God, they were gone but we can never forget "the night of the spiders."

OUR GERMAN SHEPHERD, REX

Since we now spent so much time in the bush, we decided to get a dog to go with us for alerting us to danger; a young German shepherd we named Rex. We picked him out of a litter of twelve from a breeder about three hours away where we went for supplies. He was to be our watchdog. The day we picked him up was special because it was an emotional connection that we didn't expect. Our driver and Loren were in the front seat and I sat in the back with Rex. He was snuggled up on my lap but to set a mood, I asked Loren to turn on the portable cd player with a praise and worship tape we kept going almost all the time. I loved to worship and sing but I was thinking about Rex just leaving his family of fourteen and wanted to re-assure him he was going with good people and an awesome God. When the music came on there was a presence of the Lord in the vehicle and Rex sat up on my lap and laid his face right in the crook of my neck. We prayed and thanked the Lord for the blessing of this dog. He was calm and fit in immediately.

We got Rex as a puppy recently weaned from his mama. We wanted to raise him as a guard dog and to have as a friend. At night, while Rex was still young, he would insist on sleeping in the tent with us. He was not a good bedfellow. It was a real fight to keep him out of the bed, but he was not one to give up easily as he loved us so much.

We had tried many types of beds to use in the bush up to this time, from sleeping bags to air mattresses. Invariably the air mattresses would lose air and we ended up on the ground by morning. We finally figured it out and brought a real mattress from town. We laid it on the floor of the tent. Since we had our big truck, it was easy to slide in and store and carry. It made such a great difference on how well we could sleep in this harsh environment.

Africans have a completely different concept of having a dog than Westerners. As Americans, we have dogs as pets and treat them well. In Africa, a dog is pretty much on his own to fend for himself, even if it has an owner, including finding it's own food. We raised Rex the American way. Our African team thought we were crazy. They couldn't believe we would love and feed a dog and care for him like one of the team. I reminded them we took care of their needs and provided for them because they were working on the team. Rex was no different; he was a working member of the team.

One afternoon, we heard a ruckus going on outside the tent. We found Rex, about three months old, fighting with our supper, a big white rooster that we had tied up to a bush with a string on his leg. Young Rex sized him up and thought he could take him. He was astonished at the fight the rooster put up. It looked to us that the rooster was whipping him. Although he was still a puppy, you could tell he couldn't believe how tough that rooster was. Rex brought us a lot of pleasure and comfort and some sense of being home, no matter where we were. The team soon learned to love him. He would be riding in our truck with them most of the time. His job was to fend off thieves and he became good at it.

THE VILLAGE OF KOSETEI

After consulting with the elders, we were asked to build a church in a small Pokot village called Kosetei, deep in the interior. They told us this village was spiritually a hard place and needed the Gospel. We took a team in to evangelize hut to hut and built them a church. Unknown to us, there was a large Religious compound already there. They provided medical aid and water for those who converted. We had been shockingly aware in our last ten years of ministry in Tanzania that these places existed among primitive tribes deep in the interior. Now, when the time came to preach and dedicate the new church publically, the priests stood behind the wire fence of their compound glaring and trying to intimidate us. Loren preached the Gospel and later we gave out maize to all the villagers without any conditions attached. Our little church at Kosetei struggled for some time because of the problem there. The priests told the people if they came to our church, they wouldn't get any beads, maize, water or medical help from them. It was difficult and hard on the people. Eventually these priests returned to Italy for a time and our church began to grow.

We prayed for them to be strong in the Lord and to be able to stand strong against the intimidation of being left out of getting food and water thru these controlling people. It was our first encounter with this challenge to the lordship of Christ in religious form.

RELATIVES

One evening, the principle of the Pokot School, who was a Pokot himself, came to visit us. He was a young man who had worked hard to get an education and now was back with his tribe as the headmaster of the school. He was determined to help other Pokot children get an education. We had gained his confidence and trust and he asked us if we would like to know more of the oral history of the Pokot, as they had no written history. He told us the Pokot had watched us closely and were amazed none of us had been bitten by puff adder snakes or gotten sick. He said everyone noticed every time we came to stay with them, it rained, and they considered this a blessing as I noted earlier. We were astonished at the story he began to tell us.

Sitting around the camp one evening, he told us the Pokot had originally come from Egypt, and in the flickering light of our campfire, shared some of their folklore with us. He told a story of a big dance that happened a long time ago. A handsome giant, a stranger from a distant land, attended. At the dance were three Pokot women who were all attracted to him because he could dance so well. He noticed their attention to him and invited them to come to his home, even though it was a long distance from where they lived. The ladies were excited to be invited as his guest at a banquet, and happily obliged. After finally arriving in his land, they were struck with awe at his fabulous home. They settled into their guesthouse waiting for the banquet.

Out of curiosity, one of the women, unnoticed, wandered into the big house. She was shocked to see this handsome giant eating a human being. She was terrified to learn he was a cannibal. She quickly fled, and on her way out of his country, told the other two women what she had seen, but they wouldn't believe her. The second woman decided to see for herself and sneaked in his house. She saw a horrific scene of cannibalism and she too left hurriedly, warning the third woman. Again, the third woman didn't believe her story either.

Finally, the third woman stayed and ended up marrying the handsome giant. One day when her husband came home, he tried to kill her. In her dismay she cried out, "No; instead, I will have a son and you can eat him." He finally agreed and this bought her more time. She did have a son and he began to grow. She noticed how her husband eyed him, and knew she only had a short time to save the boy and herself. Then one day, a crow came to her. The crow told her she must leave or she and her son would be eaten for sure. The crow told her to cross the Red Sea and she would be safe.

One day, when her husband wasn't home, she took her son and ran away toward the Red Sea. When she got to the water she noticed her husband, the giant, and his friends were coming after her. She was trapped with the sea in front of her and the giant pursuing her from behind. Suddenly, a powerful wind blew and opened the Red Sea and she and her son crossed safely. When the giant and his friends tried to cross after them, the sea suddenly closed up on them and they all drowned.

As we sat in the moonlight and listened to this story, we were astonished to realize how much this paralleled the Bible story of the children of Israel escaping Egypt through the Red Sea. Could it be the Pokot were ancestors of the ancient Israelites?

The headmaster of the school continued. He shared with us the Pokot custom of putting the blood of goats and sheep on the doorposts of their huts to ward off evil spirits. Again, we were amazed at how this paralleled what the children of Israel did when they came out of Egyptian bondage on the night of the Passover.

Another story he told was one that coincided with the deliverance of the Israelites from Egypt. He said they had an ancient Pokot hero, whose name was "Chimosa," who was their champion. We wondered if this "Chimosa" could be Moses because of the Red Sea story.

In Pokot folklore, he went on to say, there is a story that a long time ago one of their old men took his son to the mountain top to offer him as a sacrifice. Pokot believe God is in the mountain. Again, this reminded us of how Moses talked to God on Mt. Sinai. The teacher went on to tell us that before he killed his son and offered him, he found a dead buffalo nearby and used it as the sacrifice instead of his son. This story sounded like Abraham and Isaac on the mountain.

When a Pokot did something evil, he would slay a goat or a sheep and put its blood on his legs and fast for days to be cleansed. Another interesting thing was the Pokot don't eat meat and drink milk in the same meal as told in Leviticus under the Levitical dietary laws. We were taught Pokot also offer burnt offerings. Amazingly, these stories and customs sounded parallel to the stories in the Old Testament about the Israelites and Jewish customs.

He told another interesting Pokot story about a hyena that thought the moon was made of fat. He got the other hyenas to climb on top of each other to make a great pyramid so they could reach the moon and eat the fat. However, there came dissention among the hyenas because they all wanted to be on the top to reach the moon first and get the biggest portion of the fat. They got so angry at each other they fell to the ground. After this, they were so mad they couldn't communicate with each other. He said the fall hurt the hyena's back legs, and this is the reason why hyenas limp. This story resembles the story of the Tower of Babel. "And they said to one another, Go to, let us build us a city and a tower, whose tower may reach unto heaven; …"Genesis 11:4.

There are other interesting folk practices among the Pokot. Expectant women sing about going to the Red Sea to be cleansed. They also worshipped the sun, the moon, and their cows, exactly as the ancient Baal worshippers of Egypt.

We listened in wonderment as he continued to speak about their folklore. We had heard of the nation of Israel accepting Ethiopians who were able to trace their lineage back to Jewish ancestry. Instead of wandering into the land of Palestine, evidently these descendants of the Jews turned south after the Exodus. We knew the Jewish people had been scattered to the ends of the earth. According to the oral history, Pokot say they then turned south into Sudan. It is here, where they stopped long enough to practice circumcision after thirty years of wandering. Now, instead of only circumcising men they began to also circumcise the women. Some Pokot elders wear a permanent mud cap that sits on the back of their heads signifying circumcision and that a man can now marry. The mud cap looks like a Jewish Kippah, or skullcap (yarmulke). The use of dowries to obtain wives is another practice of the Old Testament Israelites. From Sudan, they trekked southward until they came to the land where they now dwell in Northern Kenya. When they arrived here, the land was fertile, but later it became a harsh and barren land. The Pokot have an ancient prophecy predicting their deliverer would be born of a virgin and come out of the East. Of course, this clearly points to Jesus Christ of Nazareth.

We could hardly believe what we heard. Could they be a remnant of one of the tribes of Israel? Here was a tribe where few had ever heard the Gospel, yet they knew so much in their folklore, which sounded like distorted Bible history. It made sense the further away, time wise, they moved from the Tower of Babel, the more confused the memories became. One reason they were fascinated by us was that when we spoke out of our Bible, it held so many of these stories and they wondered at how we could know their "secrets".

A VIOLENT TRIBE

We met almost nightly now to hear these fascinating oral stories of their history. Our host continued. The Pokot can be violent and are isolationists. They refuse to mingle with any other tribe. At this time, they did not acknowledge they were even a part of Kenya. Often they would say, "Kenya begins at the tarmac," and wave their hand toward the south, "very far away." They called where they lived "Pokot land." The women

192

spend most of the day walking great distances looking for water, which they fetch and carry home in five gallon yellow jerry cans. Sometimes they carry two at a time strapped with a rope to their forehead, the jugs hanging on their back and walking hunched over. They also rob honey from the killer beehives. No wonder they are so tough. The men and boys herd the cows, camel, sheep, and goats and defend them from raiders. Observing the Pokot externally, you might think they are a simple people, but in reality they have a complex and sophisticated society and culture. This is true of all African tribes.

The Pokot were infamous for being treacherous. They would raid the neighboring tribes of Turkana, Maraqwuet, Samburu, and also the Karamajong, who are located along the border of Uganda. The purpose of the raids is to steal cows and goats to increase their wealth and buy brides. The fathers of the young men won't share their cows with their sons. The quest is always to want more cows to buy themselves more wives and increase their stature in the society. This forces the young men to go on raids to get their own dowries to buy their own brides. This is the lifestyle; raid the other tribes, possibly killing men, women, and children indiscriminately during the skirmish. The other tribe would reciprocate with raids avenging themselves. This is a vicious cycle that has happened one generation after another. Time has stood still here.

It was dangerous and a big step of faith for us to be up here. We never knew at the time that one day we would be caught in one of these terrible raids. Witchdoctors held power over the villages with their witchcraft magic, causing the people to live in fear. Medicine men sold their native concoctions in the open market place. Witchcraft charms were sold to ward off evil spirits from these superstitious people.

TOM COLLINS

A missionary named Tom Collins came to Pokot in 1934, but we were told he never had any converts. The people told him they didn't believe in the white man's religion and didn't want their God. One day the Turkana came into Pokot land to raid and Collins' son was killed. We harvested the seed another missionary planted. They were unheralded great heroes of the faith. We thank God His Word is true and will not return void. Isaiah 55:11: "one plants and another waters, but God gives the increase." 1 Corinthians 3:6-9.

In the 1950's, a false prophet rose up among the Pokot. He was the leader of a cult called Mt. Zion. The Pokot at this time were desperate to rid the land of the white colonists who oppressed them. The cult leader had no problem stirring up the Pokot to fight. He told the people if they would rub his magic potion on their bodies, the colonist bullets could not penetrate them or kill them. The Pokot believed him and went to war against the colonists. Because of the promise they were given by their "prophet," they attacked the colonists out in the open, believing they were invincible. As a result, three thousand Pokot men were killed in one day. As you can imagine, this left a bitter taste in the mouths of the Pokot community.

MUSA

One of our Pokot pastors was a man named Musa. Early on as a dirt-poor evangelist, he and the schoolteacher, before he was headmaster, and other Pokot evangelist would walk village to village preaching the Gospel. Once Musa had to cross over a mountain range and the climb was steep. He had sought the Lord for the baptism of the Holy Ghost having read in the Bible where it says you would be endued with power to be a witness. "And, behold, I send the promise of my Father upon you: but tarry ye in the city of Jerusalem, until ye be endued with the power from on high." This was Jesus speaking. Luke 24:49.

While making this tough climb, he laid spread eagle and exhausted on the side of a cliff. He told us he prayed for the Lord to help him survive this climb and he was instantly baptized in the Holy Ghost and he began to speak in other tongues right on the side of that mountain cliff. He also had the physical power to finish the climb to safety. The Holy Spirit is the third member of the Godhead. The tongues are his voice and He gives us power to be witnesses. In Acts 19:2 -6. Paul asked Apollos "Have ye received the Holy Ghost since ye believed?" It is Jesus who said in John 14:26 that when He left us from the cross, the Holy Ghost would come to be our "comforter" and helper. Thank God for His provision for us after Jesus accomplished the finished work of Redemption. "...he said, It is finished..." John 19:30.

About once a month, someone was bitten by the deadly puff adder snake, a short and fat venomous viper. Miraculously, Musa said, none of them taking the gospel ever died from those snake bites. We thank God for these young evangelists who hazard their lives to preach the gospel to their people.

The Pokot had now fully accepted us and the elders of the tribe gave us a plot to camp on and invited us to build a house and live among them. We were highly honored. They brought the tribe together and hacked out an airstrip for us using *pangas* (machetes). It was one kilometer long and about forty yards wide. By hand and muscle they were able to make it level. It was truly an amazing accomplishment. The first time we flew our plane in to land at the new airstrip; we looked down and could see great lines of Pokot running from the village of Ngingyang and the bush out to the airstrip to meet us. Several hundred Pokot greeted us. It was exhilarating as we landed on the dirt strip to know these precious ones were receptive to the word of God; it looked like a dust storm billowed behind us. It was a joy to see the people coming out of the bush from everywhere to greet us. As we got out of the plane, they gathered around us singing their tribal songs and dancing. It was one of the most wonderful experiences of welcome we have ever had and it bonded us even stronger to the tribe.

We sent out our team to build a church in the northern part of Pokot in a very far village called Natan. Our cook was at the campsite for the builders when a poisonous snake about seven feet long came into the camp. It caused quite a stir and the men were finally able to kill it. We heard stories that in Pokot country, there were pythons up to thirty feet long with heads as big as cows, but thank goodness we never saw one. We don't know if this was their way of having fun with us, but we were always on the lookout anyway. This was our life in the bush. With stories going around the campfire like this, how could we miss TV?

TURKANA

During the famine, we also went from the Pokot up north to preach to the Turkana and take them food as well. It was a treacherous trip because the Turkana and Pokot hated each other so much. The Pokot had once killed an Indian trader for taking goods to the Turkana through their land. They shot out the tires of his truck, and then slit his throat. We explained to the Pokot that God had called us to help all people, even Turkana, and if they all put their trust in Christ, they could live together as neighbors. God put the faith and courage into our hearts to travel through Pokot country to go on into Turkana land despite what had happened to the Indian. We prayed the Pokot would begin to understand the love of God for all people.

As we traveled through the bush and across the dried up riverbeds, we came upon many skeletons of cows and goats but finally safely passed into Turkana land. At our first meeting, the Turkana were skittish and didn't trust us. The Turkana chiefs and elders interrogated us about why we were there. We explained to them we had come to preach the Gospel and had brought maize because we knew they were hungry. Finally satisfied, they allowed us to set up our camp. We didn't understand the grave danger we were in at the time.

The Turkana in the interior are also a primitive people. The women's necklaces were different from the Pokot and looked like dozens of chokers piled on top of each other, giving their necks a long and regal look. Their heads were partially shaved except for the top. Their clothes were also different from the Pokot in that they wore flowing robes. This was also an arid region but more so than Pokot. We pulled down the platform from off the side of our truck and spoke to them of Jesus Christ who came for all people. I greeted the people and sang songs about the blood of Jesus and Loren preached about God's love in sending His own Son to earth to save men of all tribes. "for thou was slain, and didst purchase unto God with thy blood (men) of every tribe, and tongue, and people, and nation, and madest them (to be) unto our God". Revelation 5:9b. The people responded to the Gospel and many chose to accept Christ as their Savior. We prayed for the sick while the people stood listening in the crowd. God touched these precious ones. A number of men and women testified of receiving their sight and being healed of various forms of handicap. I would always try to keep the video camera rolling of these wonderful testimonies. I felt that we were "filming good news" for what the Lord did.

Chapter 10

A Deadly Raid

We began seriously working to build churches in another area to the south of Pokot. Lake Baringo is full of crocodiles and hippos, fish and fishermen all competing for their subsistence. Our team set up a tent camp at the edge of the lake. One night, our cook slept too close to the tent fabric and was awakened by something cold touching his head from the outside. Then he heard heavy breathing. He froze, realizing it was a hippo. He thought to himself, *O God, I don't want to die this way. I have a wife and a baby daughter. What will they do without me? I'm too young to die.* Two of our other team members, who also slept in the tent, woke up and heard the quiet munching outside. One of them peeked out the tent screen and saw the huge hippo. He was about to cry out but another one of the guys put his hand over his mouth knowing if he made a sound, they would all be steamrolled. Hippos are extremely dangerous. We have been told on several occasions you can be killed inadvertently if you get too close. They all held their emotions until the giant monster moved away from the tent, but that hippo kiss became famous and we were all able to laugh about it later.

The next day, we walked out onto a small pier made of stones to take pictures of the lake. I was walking behind Loren about six feet and started calling to him to watch closely where he was going but he was so enthralled with the beauty and uniqueness of the situation, he didn't understand what I was saying. Finally I got his attention and pointed ahead to a big crocodile on the rocks about ten feet away. He was walking right up on it and never saw him. The large croc was sunning itself and was the same color as the rock. He blended in to the surroundings perfectly. Loren was almost lunch that day.

KOKWA ISLAND

We got ready to build a church on Kokwa Island in the middle of the lake. The following morning, we hired some small local boats and loaded the lumber onto them. Boats were the only way we could get across. The water was grey and we had to pass a number of hippos along the way and some of the hippos aggressively challenged our little boat. We prayed, "Lord, please don't let this engine stop," and were able to get out of their way.

Kokwa Island had several hundred people living on it and never had a church before. When we arrived we expected to see the chief or the elders there to greet us, but surprisingly we received a mixed welcome. Not everyone was happy we came to build a church. Several elders and local men of the island had stopped the construction. We had to send word back to the mainland to petition the chief and the authorities to intervene. They had already given permission to build a church there, but despite this, there was a dispute on the island as to the owner-ship of the plot. Finally, through perseverance, patience and the hand of God, we prevailed and were allowed to continue building the church.

THE KERIO VALLEY

We planned to go into the Kerio Valley to build eleven more churches and expanded from our Pokot base over the mountain into the Rift Valley. We flew our plane over and discovered the airstrip was situated in a box canyon. Flying into a situation like this, you had to get it right the first time. There could be no fly around if the landing didn't go right. The team set up our tented camp in a small village town nearby and prepared to build. This is another harsh area, not only because it gets caught in the crossfire of raids, but because of nature itself. There is massive erosion because of the powerful flash floods that periodically roar through. We had been here several times before, but on this occasion Loren went without me. I had stayed in Nakuru to minister in a special ladies meeting.

Loren shared a tent with our building foreman. The man threw his mattress on the floor on the other side of the tent and settled in. He had hardly gotten to sleep when he felt something move underneath him so he turned and adjusted himself. Whatever it was, it responded and adjusted itself, too. This went on for hours, and at one point he got up and looked under the mattress to see what it was, but saw nothing. In the morning, they discovered who the visitor was: a cobra. It crawled out from under the tent and caused quite a stir in the camp. Kenya had strict no hunting and gun laws so Loren wasn't able to carry a weapon out here. Our men took sticks and rocks and killed it. What a way to spend the night and start another day. At least it woke them up completely. Loren said he was glad I wasn't there. I had petitioned the Lord early in our ministry not to "see" the bad critters out there.

The people in the Kerio Valley were also poor and had no hope of ever having a church to worship in. The team built eleven churches in their area and installed pastors to shepherd them. Great numbers of people attended as they dedicated each church site and worshipped the risen Christ together.

TRICKY TAKEOFF

As Loren got ready to fly out of the Valley, he was concerned about how short the airstrip was, so decided to fly alone to keep the weight down. Since the airstrip was in a box canyon, there was only one way to take off. Fortunately, the direction of the wind was in his favor. There was a dirt road that went by close to the takeoff threshold and it intercepted the airstrip at about a forty-five degree angle. He decided to use the road to help him build up airspeed. The team watched the road to make sure no one would drive into the plane or any animals or people would wander onto the runway. He taxied down the road a little ways. After applying the brakes and giving it full power, he released the brakes and began to roll. When he intercepted the airstrip, he doglegged with a hard right turn onto the airstrip. The turn was so hard that the left wheel lifted off the ground but then once on the airstrip, came down again. He let the airspeed build up until he was nearly at the end of the runway, and then pulled up. The plane flew beautifully, but he said his heart was pounding. This innovation in taking off entertained our team. Later he rendez-voused with them and met me back at our base camp. He loved to tell flying stories.

SAMBURU COUNTRY

This was a six month tour and we turned our attention now to work in Central Kenya building churches at Wamba and Archer's Post. We flew over this wild terrain and coordinated with our building teams so they could get started ahead of us in the vehicles. With the plane we increased our productivity. Many areas that took hours or days for us to get there by road could now be easily accessed in a fraction of the time. We could get the teams started in several different places at one time and then go in by air to evangelize. Heading out to Samburu, we had to climb over the mountains and across a great wilder-ness. We would see an occasional *manyatta*, a Masaai family dwelling, scattered out sporadically along the way. They were distinguishable from the air by their circular thorny brush fences called Kraals. We always carried survival gear, extra dried food, and water in our plane in case we had to make an emergency landing. Flying by GPS helped us navigate, reaching into far and remote places with pinpoint accuracy.

Loren gritted his teeth when he saw how the airstrip at Wamba was laid out. When we say, "airstrip," we are usually talking about a dirt landing strip cleared of bushes and trees. On the right side of the airstrip was a mountain and at the other end of it, maybe a hundred yards, was another mountain giving it an L shape. The direction the wind blowing made us have to fly close to both mountains so Loren could properly setup to land into the wind. I always prayed for us and today was no exception. Because the threshold was so close to the mountain, we had to make a steep angled decent. We landed safely and the Samburu people came out to see who we were. Samburu are a sub-tribe of the Masaai and we had been warned they were fierce as well. Maybe it is because they live so isolated and see so few foreigners. Naturally they would be wary of strangers. They were not sociable at all. Even the women looked as if they could take a lion down with a stick. We hired a night watchman to keep people away from the plane and the next morning, he told us he had hyenas around the camp all night long.

The Samburu people appeared taller and bigger than their fellow Masaai of the Mara. The Moran are the young men from middle teens up to thirty-five. They are the warriors of the tribe whose job is to fight off raiders trying to steal their cows, keep the village safe, and raid other tribes themselves. The Moran are not allowed to get married until after thirty-five when their Moran days are over. This means the younger women are married off to older men. The Samburu Moran are very menacing in appearance. Many of them wear feathers in their long braided red wigs and wear the red tribal dress. They were nothing to tangle with. Not only did they have spears, native bows and arrows, but also some antiquated guns. These were probably some of the fiercest people we had encountered to date based on appearance.

BISHOP ARRESTED

We pre- fabricated as much of the church buildings as possible and our trucks transported them in this way plus the rest of the lumber we would need. The overseer went ahead of us to prepare the building sites. By the time we arrived and started building, the bishop had gotten arrested. Wamba is a major center for the religious organization we spoke of before. In the middle of the most remote places they would go in and tie up the inhabitants with a mixture of Christianity and witchcraft. They did not like the idea of us building a protestant church there. Unbeknownst to us, a woman offered our contact the use of her car, a very rare thing to have a vehicle out there because of the remoteness. His vehicle had broken down. Our Bishop thanked her and after he began using it, she went to the area police and accused him of stealing it. He was in and out of jail the whole three month's we were there building and preaching. It was evident he had been intentionally set up to delay or even stop the building of the church. It was a terrible time of tribulation for him, but eventually he was exonerated. We thank the Lord he was strong in faith and never complained. These types of problems unfortunately are part of what we sometimes deal with. We built wonderful churches in Wamba and Archer's Post, but it didn't come easy.

Most of the time, the villagers would show their appreciation for our work by giving the team a goat. They are usually good meat and appreciated since we have many mouths to feed, but in Wamba we got an old Billy goat. Our cooks made the usual stew out of him, but the taste was so strong, it nearly crossed our eyes. Whatever that goat ate affected the taste of the meat. This one we had to say grace over twice but nevertheless we were truly thankful to the community for their show of appreciation.

The next day, we got up planning to go to Archer's Post to dedicate the other church. We got in the plane and headed to rendezvous with our team there but we ran into a bad storm and had to turn back causing our ground team to do the dedication. We had no consistent reliable weather forecasting. We had to learn the weather patterns, react to what we ran into, have an alternate plan, and be cautious.

The return to our base camp turned out to be a great tribulation for the team. The dirt road was so bad they had eight "punctures" (flat tires). Breakdowns are interesting in that if you come upon someone broken down, your first clue is you start to see branches strewn along the road. Soon, you come upon a stranded vehicle and then beyond it, the branches are again strewn ahead for some distance to warn anyone coming in the opposite direction. For our team, this return trip was a constant litter of branches times eight. It took a long time to break the tire off the rim from our seven-ton truck. We later were able to replace our wheel rims with Mercedes-Benz rims. This stopped a lot of our tire-changing problems. In Africa, there was a shortage of original parts and the substitute ones can cause constant problems and most of the time we had no choice of availability of any part. Our German Shepherd, Rex, loved to travel with the team and he was the protector of the truck. He knew his job and he loved it. At some points when the team broke down, thugs appeared to see what they could steal, but Rex kept them at bay. His size and bark put fear and intimidation into anyone trying to approach the team or our trucks.

TURKANA RAIDS

The Lord put it on our hearts to return to Turkana country to evangelize deeper into unreached villages and to continue to build churches there. Turkana is some of the most desolate and forbidden country on this planet. Much of it is solid sand and dunes, except for an occasional oasis with palm trees along riverbeds that were most often dry. In looking out over the vast expanse of sand, at times with the reflection of the sun, it produced a mirage of lakes of cool water.

Turkana houses look like cone shaped birds' nests and were made out of sticks. Many of them were on stilts to keep them above the occasional flash floods, the many snakes, and mostly the heat radiating from the sand. The Turkana, as we mentioned, in the northern interior are also primitive people. Before they marry, the male suitor and his bride to be must fight each other with long sticks. If the woman wins, she doesn't have to marry him if she chooses and the family can keep the dowry he brought. If she decides to marry him after winning the fight, she will be the boss of the family. So the fight, as you can imagine, is vicious. Turkana women are tough. Like the Pokot, they spend much of their day searching for water and food and in general fighting to stay alive.

We sent the team ahead to setup camp and start the building process while we checked on other areas we had been developing. In time, we flew over the vast wilderness to Lochichar and were greeted by our team at the airstrip. We hired local armed guards to watch over the plane. It was hot, even late in the afternoon, and we soon realized the heat would not set with the sun. We went to our tented camp and I began to setup housekeeping. No matter where we were, I made things as nice as possible. This was our home no matter where it was and I wanted us to have some consistency if it was only where we slept. We had two cooks with us whose main job was to feed the team and keep the clothes reasonably clean. Their job is tough and starts before sunrise and lasts up to ten or eleven at night. We always hired extra locals to help them because often we fed up to thirty people a day, depending on the area and how many extra hands we had to help build the churches.

THE DEADLY RAID

The next morning at 4:00 A.M. while we slept in the main camp, a few kilometers away at one of our building sites, five hundred Pokot came in with a massive raid on the village. They were armed with AK-47's and G3 African military rifles. They killed everything in sight: old women, old men, women, children, and even babies. Then they burned down the huts and stole the cows, goats, and camels.

When we heard what had happened, we immediately sent word for our building team to return to our main camp if they were alive. The team never responded and we were quite worried about them. About six hours later, we learned what had actually happened and why they didn't come back. They sent word they decided they would not let the devil win and take this village to hell. They would not stop building the church to run away. They continued working on the church, hardly able to distinguish the hammering from the gunfire. The Pokot knew we were in the area and this was one of our churches in the sense that we were building it for Turkana. They were overheard yelling, "Eat Davis' churches; we're taking the cows." By this time, they had taken hundreds of cows, camels, goats, and sheep.

We had originally set up our campsite on the banks of a beautiful tree-lined riverbed that had a little water in it and lots of shade. We thought this would be a safe and cool place, but soon discovered it was full of long thorns and were constantly stuck through our sandals and boots. In the heat of the day, we sat with our feet in buckets of river water and tried to cool down as much as possible. Loren tried to be innovative in keeping me as comfortable as possible.

When we were told what the Pokot raiders yelled out about us during their raid, it infuriated Loren. We began to cry out to the Lord to not let them get away with the stealing and killing. Then the Turkana men with their G3 guns began to counterattack and go after the Pokot. We were advised to move our camp to try to find a more secure area, although no place was absolutely safe.

Moving hundreds of cattle, goats, sheep, and camels is no quick task. The battle raged between the Pokot and Turkana for about three days. We never considered leaving and our teams continued building. It was by the hand of God none of us or our team were killed. During the hit and run pitched battles, we drove our truck to the ravaged village to dedicate their new church. Our team had finished the building and the people were ready to have a service in the middle of this chaos. We had to go through a thick jungle area on twisting, winding paths to get to this particular village where the killings had taken place. When it was said and done, eighteen villagers lay strewn throughout the village, slaughtered. Believe me, we prayed all the way because we never knew if the Pokot raiders would meet us at some turn on the jungle road and shoot us. We did meet a Turkana militia on the way and we stopped and prayed with them. They held their guns with bowed head as we prayed for safety for everyone involved. We had armed guards traveling with us, but still it was dangerous. Returning from the dedication, a bad storm hit and it was a great struggle to keep the truck from getting stuck in the sand. A big tree was knocked down across the dirt road by lightning and we had to use our chain saw to cut it apart so our truck could pass.

The Lord helped us and we completed four churches in that region. Several Pokot raiders were killed and after all was said and done, they only got away with a few cows in all the commotion. Back in Pokot country, we later found out the elders disciplined the young raiders because it was an unsanctioned raid. Some of those raiders were made outcasts from the community over this disobedient act. When we flew out of our Lochichar camp, we climbed in a tight circle until we got over a mile high. A military pilot had once told Loren if we were ever in danger of being shot at from the ground, to fly at least a mile high and not to fly in a straight line but an evasive zig zag. Normally when we flew over Pokot, we would fly fairly low so our friends could see us and meet us at the airstrip or could see we were on our way somewhere else. At least they knew we were around. This time we flew over Ngingyang at a high altitude not trusting what their attitude would be after this recent skirmish. Later some of the Pokot sent word wondering why we flew over so high. We had to stay alert and not be foolish. Wisdom is the principal thing. We weren't about to take a chance until we could talk to them in person.

Chapter 11
Crowns and Storms

We came in after an extended time in the bush to the farm base we used thru the generosity of a businessman outside of Nakuru. Almost immediately after arriving, a friend of ours rushed over excitedly. He said, "You have been called to the Statehouse to see the President of Kenya. You are expected to be there in three hours."

Astonished and unprepared, we ran around frantically, trying to find our best clothes, get them pressed and get some water heated for baths. I was in a real panic because we had been out in the bush so long and my hair was a mess. After a few minutes I decided to wash it and start over, but how to get it dry was the problem. We didn't have enough time to let my hair dry naturally and of course there was no hair dryer or electricity. I wasn't about to go to see the President of Kenya not looking my best. Dejectedly, I told Loren to go on without me. He told me there was no way he would do that; we were a team and that I needed to find a way to pull myself together. A half hour later he found me drying my hair in the hot smoky exhaust of the diesel generator. Loren was so amused and thought that was real entertainment, but had to keep his amusement under control because I was so stressed about how this was going to turn out. I got all dressed up, sprayed my hair with perfume to cover the exhaust smell and an hour later we were on our way to the Statehouse. Loren said I looked great and no one would have ever known we had come from the bush and what I had to do to look like I did. The President was so gracious to us. He had heard about our work among the Pokot and how the Gospel had impacted them. The reports of the overall raiding had decreased in the area and he wanted to meet us and know who we were personally. We never had any idea what the Lord was doing through us deep in the bush or that it would ever be noticed and recognized by the highest office of Kenya.

MOMBASA, KENYA CRUSADE

We headed to the east coast of Kenya to the city of Mombasa on the Indian Ocean. The entire coast of Kenya is predominantly Islamic and Mombasa was a major city. We now had a highly experienced advance man who worked with another major crusade evangelist and he wanted to help us. We advertised for the crusade more than we had ever done before. Early on, we picked up there were grumblings from some of the pastors who wanted us to give them a big sum of money in exchange for cooperating with us. Unfortunately, this problem was spawned by many *wazungu*, western preachers, who had preceded us in Africa. They would give the pastors money in exchange for helping them. The product of what they had done was the corruption of so many African pastors. They did not want to do anything, even for the Lord, without getting paid for it. We were sure the early whites had felt sorry for the "poor" Africans and thought they did a good thing, but now many of the African pastors had perfected the art of extortion on visiting evangelists and missionaries from abroad. Our experience in Tanzania had shown us Africans would take up an offering for a visiting African minister, but not for whites. It had nothing to do with the Gospel. They considered everyone with white skin as rich. Our policy was our crusades were a cooperative effort and the pastors were the ones who would benefit by having new believers in the church. We always tried to be fair and appreciate what anyone did by giving them a "blessing," but pastors wanted to be "paid" large sums. It was a real problem and gave us more headaches and heartaches recognizing where they were spiritually.

In spite of this inner battle with the pastors, Loren pulled a banner over Mombasa for three days to advertise the crusade. One of the rules for this procedure is you have to stay at least a thousand feet above the highest obstacle. He pulled the banner all over Mombasa and word came back the Muslims were saying, "The Christians have been coming by land. Now they are coming by air. I guess next they will be coming by sea." They were shaken. No one had ever seen a banner pulled behind a plane before. Sometime later that year, it was reported that Al Qaeda shot a Stinger missile at an Israeli jet over the city of Mombasa. At the same time they also bombed a couple of tourist hotels. After that incident, Loren decided not to pull a banner over Mombasa again.

The crusade started and the Spirit of God began to move. Christ was exalted and many mighty miracles happened. One of them was a blind woman who was led to the crusade by a six-teen-year-old teenager. She received her sight and demonstrated it by counting Loren's fingers held up in the air. A young man who had been deaf from birth received his complete hearing and could repeat a whisper. A man who had urinated blood for ten years testified the flow of blood had completely stopped. A lady who was oppressed by the devil felt those powers broken off of her. Another lady whose breasts had been swollen badly testified she was now normal. This is one reason we know the testimonies are real because they are so personal. No westerner would confess to some of the things we have heard. The Lord manifested Himself in Mombasa.

In the mornings during the crusade, we held a pastors conference. Every wind and doctrine had come to Africa from Europe, the UK, and America and there was much doctrinal confusion. Overall, the pastors rebelled against the call back to biblical teachings, and forbid any of their church members to attend these meetings. They didn't want them to hear any truth and stop their control over the congregations. They became hostile, especially when we dealt with the abuse of the prosperity message. The problem was much deeper than we had imagined.

Despite the rebellion among the pastors, God still moved mightily in the crusade. We had a tremendous crowd on the last day. Despite that, we found out those pastors planned to take the microphone from Loren that day. The crowd was so large people hung from trees and stood on buildings all around the field. There was a hunger to hear the *real* Word of God. It was amazing and wonderful that so many people wanted to hear and understand about Jesus Christ.

A CHURCH FOR ZANZIBAR

Our team broke down the equipment from the crusade field and packed it back into the truck preparing to leave Mombasa. We were going in separate directions and planned to rendezvous with them back at our base in the Rift Valley in a month. Loren and I and another African couple headed for Zanzibar to get a church established there. The team would be heading out in the truck for the long slow journey back to base. We always met together for a debriefing and prayer meeting after a crusade and before parting, but this time I was uncomfortable in my spirit. After the meeting and prayer, I hugged the team more than once each. As we said goodbye, I began to cry, not knowing why.

We flew to Dar es Salaam and from there took a hydroplane ferry over to the Muslim island of Zanzibar. It was about twenty miles off the coast of Tanzania. At that time, Zanzibar was ninety-nine percent Islamic and had once been ruled by the Sultan of Oman. Tanzania was still known as Tanganyika until it got its independence in 1962 and joined with Zanzibar to become one country, renamed Tanzania. "Tan" for Tanganyika and "Zan" for Zanzibar. Unknown to us at the time, Zanzibar was also a hotbed for Al Qaeda. Thinking back, we recalled the entire belly of the hydro boat was full of Tanzanian soldiers that day we went over. The passengers had to stand aside when we docked to let the soldiers disembark first. We had no idea what was going on.

On earlier visits to Tanzania we had heard stories of a brave pastor working on Zanzibar and we felt in our spirit this would be the man to work with on our mission. He had come from Kigoma in western Tanzania where he had been a successful businessman, but left his business because he felt the call of God to come to this rock hard mission field. By now, he had been in Zanzibar for seven years, but had seen little fruit in his ministry. He was poor and paid rent on a cinderblock house that was still under construction. We were surprised to discover it had no windows, doors, or even flooring. Furthermore, they had no furniture whatsoever. They and their ten children slept on a thin blanket covering a straw mat on the concrete sub floor. There was nothing to keep the mosquitoes out. We admired him and his wife tremendously for what they had given up and for what they had come here to do. They had committed their lives to bringing Christ to these people and living the Gospel in front of them. The Muslims had persecuted them incessantly and boycotted whatever business they tried to start to support themselves because they were Christians. We felt he had the tenacity that, if we were able to build a church here, he would be tough enough to hold onto it.

We soon realized the Muslims would never give us a permit to build a church or allow us to finish an unfinished building if we were even able to purchase it. We changed our strategy and looked around the island to find a nice size house we could convert into a church. We had $18,000 to work with. We knew if the sellers saw our white faces, they would be suspicious and never sell, so we decided to let our Kenyan team members do the looking and negotiating. We played the tourist role and took a laid back approach.

Our team found a couple of prospective houses, but always ran into a snag with the owners during negotiations. The sellers were highly suspicious of why Kenyans wanted to buy the house. They found it was unusual as no outsider was really welcome. One man said he would sell his house, but the next morning the Muslim sheik and some of the elders came to visit him. They sensed in the spirit we were Christians and would use the house for a church. We tried to be quiet and discreet but it was a small island. The sheik told the seller if he sold his house to us, he and his friends would come back and kill him. This was extremely disappointing and frustrating, but we were determined and felt strongly we were following the Lord's leading so we didn't give up looking.

Something else happened in the spirit realm. Two days after arriving in Zanzibar, we got an emergency call from our team. To our horror we found out they were still in Mombasa and had been involved in a terrible accident with the truck.

I have to explain at this point that roads in Africa at that time were for everyone from donkeys to people pulling carts to big eighteen and twenty wheel trucks. Especially in Mombasa, which is a port city; trucks transporting goods were everywhere. A man pulling a cart had changed lanes and pulled out directly in front of our Lorry with no warning. Our driver Juma swerved to avoid hitting him, but instead of the man turning away from the truck, he panicked and continued to turn right in front of it. Juma swerved hard to the left trying to miss him, but to no avail. The poor man was killed instantly. When Juma swerved hard to the left, the truck hit a mound of dirt beside the road and flipped the Lorry over, narrowly missing going over a cliff into a lagoon in the ocean below. For certain everyone would have drowned. Juma's wife Helen was trapped inside the truck, and Juma had to break a window to pull her out. She suffered a broken arm, but none of the rest of the team was seriously hurt. Our men had to stand guard over the truck or thieves would have stolen all of our crusade equipment. There is no mercy when an accident occurs here. Many poor see it as an opportunity to prosper. The accident happened at a place well known as a "black spot" in Mombasa. A "black spot" is a place where fatal accidents happen over and over in the same place. Most Kenyans believe witches are behind the accidents and are caused for the purpose of human sacrifice believing the devil requires blood.

We were in a terrible bind when the accident happened. Our team faced a major crisis and we had not finished our work in Zanzibar. We couldn't be in two places at once. Our truck had been towed to the police station where it was impounded and in bad condition. We sent word to Juma to try to repair it as much as he could, but we did not understand the gravity of the situation. The impact was so great that the front axle had sheered in two.

It was a painful dilemma to deal with. The truck accident pressed us hard to return to Mombasa as soon as possible, but we hadn't finished what we came to do in Zanzibar and our negotiating process was fragile.

After talking with the team and praying about what to do, we chose to stay in Zanzibar a few more days and keep slugging out the house church dilemma. The team felt confident to deal with their issues until we were finished. The Lord supernaturally helped us and we were finally able to purchase a big blockhouse that was completed. The negotiations were tough because the woman selling it was nervous about the sheiks and wanted it done quickly. She was a widow and wanted to get off the island and get to the mainland as soon as possible. She demanded the money before we got all the documents together and signed, but there was no way that would happen. It was a dangerous deal, but God helped us to prevail and we got great victory in the end. We registered the house in the name of a friend of ours whom we trusted and had known for years in Tanzania. He was also the Major Bishop of the denomination the church would be registered under. The deal was also Tanzanian to Tanzanian which made it tighter legally. Now the pastor and his family who had been living in such miserable conditions were able to move into the house and use part of it as their living quarters. They converted the other part of the house into a church. This was a massive victory.

Now we were free to fly back to Mombasa to deal with the situation there. After a lot of difficulty, we got the truck repaired and were able to get it and the majority of the team started on the road for our home base. An amazing fact came to light at this time. The pastors in Mombasa had actually been rejoicing over the accident saying this was the judgment of God on us for not giving them the huge sums of money they demanded from us. They said the truck would never be used again. We made sure the truck drove all around Mombasa for everyone to see before it left for our home base.

PRISON

Juma had to go to court because someone was killed in the accident. During the trial, the Muslim judge kept asking him if he had any "oranges" to give him. This was a code way of asking for a bribe, but Juma refused. The judge knew Juma was a Christian and the truck and team were part of the wonderful crusade in Mombasa. The witnesses to the accident all testified on Juma's behalf, clearly explaining the accident was unavoidable but because he wouldn't "play" and because of his faith, Juma was sentenced to three years in prison. We were all devastated. Juma was brave and took his sentence like a man while we began to fight back on his behalf.

In an African prison you can imagine the many trials he must have gone through. This is not an American prison. The prison where he was locked up didn't have clean cells nor did they have hygienic bath and toilet facilities or food and clean clothes every day. They don't have television and libraries or workshops to develop any skills. If you don't have any family members to bring you soap, you don't even have that. Juma began to use the little soap his wife brought him, cutting it in half, as a reward for any prisoners who would come hear him share the gospel.

Instead of beds and mattresses, there were thin mats on the floor to sleep on and it was so crowded the prisoners had to sleep on their sides. Juma told us if one got up in the middle of the night to relieve himself, he couldn't find a place to lay down when he got back. Homosexuality was rampant and there was little food. Juma was a stout, muscular man when he went in and lost a lot of weight on the meager prison diet, but like Joseph in the Bible, his stock began to rise in prison. Before long, he became the prison chaplain and was allowed to conduct services. During Juma's imprisonment, ninety prisoners gave their lives to the Lord.

Normally, non-family members are not allowed to visit prisoners, but because we were well known in Mombasa from the crusade, the guards graciously allowed us to see him. It did our hearts so much good to see each other and to grasp his hand through the prison fence although touching was not allowed. Juma's spirits were not down at all and he told us how God was using him there.

God touched some of our partners who felt led to help support Juma's family while he was in prison. They also gave us the funds to fight for his release. When we told Juma about this he greatly rejoiced in the Lord because, of course, he was worried how his family would make it. We all prayed and worked together to free him. After nine months of a three-year prison sentence, through the hand of God and the prayers and giving of our partners, Juma was released.

A TERRIBLE STORM

We stayed in a tourist hotel in Mombasa while wrapping things up regarding the accident. Now we looked forward to flying back to our base after making sure Juma was taken care of. We tried to relax the day before leaving and decided to try the free scuba lesson offered in the hotel pool; but I had to stop in the middle of it because I suddenly got severe claustrophobia. This had never happened to me before. I was badly shaken for the rest of the afternoon and even that night I felt "closed in" even in our hotel room. We made it through the night and the next morning got in the plane and headed back to Nairobi and on to the place where we stayed three hours away. We had been through an arduous, but victorious trip. Little did we know the most difficult trial was ahead of us.

Loren had equipped our plane with a portable oxygen tank, which would help when flying in higher altitudes. We climbed up to a good cruising altitude and leveled off. There were a few scattered clouds and it looked like a good clear flying day. As we have said before, the weather service is not reliable and soon the clouds began to close in and kept rising. We climbed to 16,500 feet and leveled off. We were now on oxygen. In what felt like no time at all the clouds totally covered us again and we were unable to climb any higher. I told Loren later that the claustrophobia came over me again, and it was so disturbing that it was only the Lord who helped me hold myself together. I knew I couldn't "lose it" for Loren's sake as a pilot. We were thankful he had gotten his instrument rating at this moment.

Suddenly, a hard wind shear tore at the plane. Shaken but each of us trying to stay calm, we made the decision to turn around and head back to Mombasa. We knew Mt. Kilimanjaro was on our left side at over nineteen thousand feet high and to our right was Mt. Kenya, which is over seventeen thousand feet. We sure didn't want to take a chance of flying into any "rocks" that were hidden in the clouds. As soon as he got the plane turned around, we heard the loudest noise imaginable. It sounded like the space shuttle at liftoff. Then a mighty force hit us from the right side at about one o'clock. It hit us with such a blow that it knocked the plane straight down toward the earth in a hard left turning motion. It was like we had been hit with a fly swatter. We were totally socked in the clouds unable to see anything. I was now fervently praying aloud in tongues.

217

Loren scanned the instruments to interpret the plane's actions. The altimeter arrow twisted rapidly to the left indicating we were in a hard dive. He knew by situational awareness where we were and that he had some altitude to work with, but his first job was to get the plane under control. Watching the artificial horizon instrument, he leveled the wings so we would be in a straight down dive instead of a twisting one. Then he pulled the throttle back to stop us from being in a power dive and then slowly and gently pulled back on the yoke until he got the plane level with the earth. He knew if he pulled back hard, it would blow the tail and wings off the plane.

I was so unnerved that my legs pounded each other I shook so hard, but to my credit, he said I never screamed. I continued to pray. When he finally got the plane level, he noticed the altimeter still spinning downward at the same pace, losing altitude in mega bites. There was nothing else he could do. Some pilots told us later we were either in a microburst or a tornado. Normally, in this type of situation, no one survives. As we rapidly descended, Loren put in enough power to put the plane at maximum turbulent speed and the stall horn screamed, warning us a stall was imminent. We were between the devil and the deep blue sea. It was awful.

All he could do was hold the plane straight and level. We both knew only God Himself could get us out of this. Then, out of nowhere, we hit the updraft and it shot the plane up like a rocket as fast as we had gone down. Loren gave a sigh of relief, because even though he had no control of our altitude, at least we went up instead of down. The clouds would be a lot softer than the ground. The plane spiraled upward through the clouds and he leveled the wings and added power, and then once again slowly leveled off. After a bit, we shot out the top of the clouds and our altitude stabilized. Soon after leveling off, another hard wind shear tore at the plane. Loren kept flying the plane at the slowest maneuverable speed to keep it from coming apart. We have no idea how many wind shears we hit. Loren relaxed and didn't fight the plane; he only did what he could.

He told me to set the GPS toward Mombasa because at this point he had no idea what direction we faced. Continuing to pray, I calmly followed his instructions. As we kept hitting wind shears, which made the plane shake so hard, I took out my Bible and started reading aloud the ninety-first Psalm. That's exactly what we needed to hear. I read down to the last verse, "With long life will I satisfy you." Thank God I kept calm and did not allow the claustrophobia to dominate me; otherwise I could have grabbed Loren or screamed uncontrollably. That would have been disastrous. Later I told Loren I not only wanted us to hear God's Word, but wanted the devil to hear it as well. It gave us both courage. We kept repeating the last verse, "with long life will I satisfy you and show you my salvation" Psalm 91:1-16.

After getting the plane under control again, Loren finally was able to call Approach Control in Mombasa and advised them of our situation and alert any other pilots flying this direction. Finally, at last, the clouds broke and we saw the emerald green waters of the Indian Ocean with Mombasa nestled on its shores. After landing, we both got out of the plane and literally kissed the ground. Our ordeal had gone on for more than an hour. Thank God we never let the fear or panic control us; it was there, but we didn't let it control us. Otherwise, for sure you wouldn't be hearing this story. There on the tarmac, we thanked God for saving our lives. We were able to pull ourselves together long enough to get to a hotel and once inside our room we completely broke down, held each other and cried, trembling violently. I told Loren as we were in that death spiral, the devil screamed in my ear, "There won't be enough left to bury you in a matchbox," but God.

The next day we had to try to fly the same route back to our base. Understanding the trauma we had gone through, Loren offered to fly me back to Nairobi on a commercial flight or send me on a bus, but I said I didn't want to leave him to fly that route again by himself. I told him if I didn't fly with him now, I would never get in the plane again. I did ask him to stay out of the clouds, which was a no-brainer. We had also met another pilot that day who told us the trick of flying in that area: stay low. We flew back, staying a thousand feet above the terrain, which kept us under the clouds. This time we had redefined IFR flying to mean "I Follow Roads." The weather was good and we had an uneventful three hour flight back to our base although we were still pretty unnerved by all the happenings of the entire Mombasa experience.

Chapter 12

Spears

We were camped in Pokot country checking on our existing churches and working on logistics to add more to this vast community. We had planted twenty-seven churches among the Pokot tribe by this time but more and more communities were asking for a church.

About seven a.m. one morning as we prepared ourselves for the day's work, one of the Pokot Bishops came running into camp. He was out of breath and calling out loudly to us and waving a short wave radio. We immediately sensed something was terribly wrong.

In great distress, he began to say over and over, *"Pole... Pole sana,"* "I'm sorry, I'm so sorry," in Swahili. We were confused. Finally he blurted out, "America has been attacked." It was September 11, 2001. Stunned at his words, we asked for details. He said two airliners had been flown into the twin tower buildings in New York City by Muslim extremists.

Our minds quickly recalled how, a few years earlier, the American embassies in Nairobi, Kenya and Dar es Salaam, Tanzania had been bombed simultaneously by Muslims. We had seen the devastation of the bombing in Nairobi with our own eyes. The American Embassy at that time was located in the downtown area. The embassy and surrounding area were blown up and so many people were injured and killed by the flying glass and debris. It looked like a war zone in Nairobi.

We were both stunned and immediately began to ask him so many questions. Since he had just heard the report, he didn't have any details. Usually over shortwave radio, the news from the U.S. at this time consisted of some short briefings, maybe ten minutes, on VOA, "Voice of America." We couldn't get many details.

We had a satellite telephone with us which we carried in the plane in case we ran into an emergency and got stranded somewhere. During those days, the satellite phones were the size of a small, heavy laptop computer. It was also helpful because with it, we could now keep in touch with our children back in America. We scrambled to set up the phone, but it wouldn't work. We couldn't get any signals from the satellites to triangulate. We reasoned that for security the satellite signal was probably blocked. After what felt like an eternity, we were finally able to get through to our kids to make sure they were okay and to get more details about what had happened.

In East Africa, we were nine hours ahead of Central U.S. time, so for our family, the horror was still fresh and unfolding. We felt so helpless and immediately began to pray and intercede for our country, not knowing what might happen next. Our family was in Texas but my brother lived in New York City and I was concerned about him because I knew his jogging path was right near the Twin Towers. The American Embassy in Kenya was warning all Americans to stay put and after talking with our family and a lot of prayer, we decided to complete our current mission.

GOING TO MAASAILAND

The Masaai are some of the proudest and most regal people in Africa. They live scattered in many different areas, but the Masaai Mara in Kenya is their most well-known traditional land. Before boundaries separated the land into countries, the Masaai land included northern Tanzania and the central Tanzanian Masaai Steppe. This is the only tribe the British colonists were not able to placate and conquer. When we came to Masaai land, few Masaai had been evangelized. Up to that point, they had been resistant to the Gospel.

The Masaai are a strong community and have their own distinct culture. The women shave their heads except after childbirth. They decorate themselves with handmade jewelry including long colorful beaded earrings and their own Masaai *shangas* made of small beads in patterns. The women make the loaf shaped houses out of mud, grass and sticks and cover it with cow dung which hardens. One entire Masaai family lives together in a grouping of houses inside a circular fenced area known as *manyattas*. The entire manyatta is enclosed with a thorny, bush fence. The center of the *manyatta* is an open area for keeping the cows and goats safe from lions and other predators at night. During the day, the livestock are taken outside the compound to graze in the wild.

A typical Masaai family consists of one husband and many wives. Each wife has her own house where she and her young children stay. The structure is relatively small with dirt floors. These are not living quarters, but temporary houses and used only to sleep and shelter them from the elements and danger. The beds are made of built up packed dirt covered with cowhide. There may be a privacy curtain for use when her husband visits. Only the smallest children sleep inside with the mother. A smaller room is set aside inside this shelter where newborn calves and goats are kept for their protection at night. In the corner is a place for a cooking fire during the rain and for keeping warm at night in the cold months. The altitude is over five thousand feet in the Mara. You can tell the number of wives in a *manyatta* by counting the houses. There are lots of children and the patriarch visits his different wives at his discretion. He has his own house.

The older male children take care of the cattle in the daytime while the younger boys are shepherds over the goats and sheep. They sleep in groups together. The Masaai traditionally don't bury their dead. They, like the Pokot, put their feeble elderly out in the bush to be eaten by lions and hyenas. Since the Gospel has come, for believers, this practice has changed.

Masaai follow a similar tradition as their Samburu cousins in their initiation rites into manhood. There is a special ceremony where young men become a *Moran*. This involves circumcision and other traditions the elders pass on to them. The *Moran* wear long braided red wigs and the distinct Masaai *shuka* , a red cloth, wrapped regally around them. The traditional *Moran's* main diet is blood mixed with milk from the cow. They get the blood by piercing the neck with a small spear into the cow's artery. Masaai do not eat wild game because they believe it will bring a curse on them and cause their cows to die. When Masaai men travel on safari, which is the Swahili word for "taking a trip" or "journey," they may stop to visit a friend's *manyatta* on their way. Tradition dictates the husband of that *manyatta* will offer the visitor one of his wives for the night as a gesture of hospitality. The male visitor puts his spear in the ground in front of the wife that he wants for the night. This is a *culturally* acceptable practice and children are looked on as members of the Masaai community and every child is accepted, no matter who the biological father is. He is Masaai. Many of these traditions are changing among Christian Masaai.

Masaai also practice the dowry system. Daughters are sold as brides to whoever will offer the most cows. It is not uncommon for an old man to buy a girl of ten to fifteen years old to be his bride. Sometimes these matches are made while the girl is a baby. The girls are often taken from their families while still children and it is traumatic. The more daughters one has, the more cows and livestock the patriarch can get from dowries.

Traditionally Masaai worship one god, Ngai which means sky and was once connected to earth but they separated. Ngai is neither male nor female but has different aspects. Maasai believe that all the cattle in the world were given to them and they need the grass of the earth to feed the cattle. They believe in death the evil person will be carried away to a desert place and the good go to a land of rich pastures and many cattle. There are other variations to the story. We have been told of a god of lightening by one elderly masaai woman who said that her god, lightening, would not allow her to accept Jesus Christ as her God. That was a heartbreaking conversation for me. She wanted to accept Jesus Christ but she feared making her "god" angry and could not be persuaded. She said she wanted her children and grandchildren to know Jesus but she was convinced she "could not".

LION HUNTERS

The Masaai Mara is one of the greatest natural places of African wildlife in the world. It is full of Elephant, Giraffe, Zebra, Rhino, Wildebeest, Hyena, Cheetah, Leopard, Lion and Buffalo, to name only a few. It is one of the most magnificent places in the world to see God's creation in the natural. The lion and the Masaai are neighbors and natural enemies. As pastoralists, the Masaai are constantly at war with the lion over their cattle. There is always a problem, but especially when the migration of wildebeest is over and there is little food on the savannah for the lions. This world famous migration starts in the Serengeti in Tanzania and makes its trek into the Mara in Kenya. The animals stay for several months until the grass is eaten up and then they migrate back across the Mara River to the Serengeti where the rains have replenished the grass there. It is this large concentration of herds which attract tourists to Tanzania and Kenya in droves. Lions do not migrate—they are territorial—so when the "buffet" has moved on, the Masaai cattle become the main source of food.

One of the rituals the Masaai *moran* must go through in entering adulthood is to kill a lion. It is their job to defend the tribe, the cattle, and other livestock against any predator. After his service to the tribe and after he has killed a lion, he is now eligible to take a bride. He is usually in his thirties and has finished his service as a *moran* when he takes a wife. The women say, "Why should I marry a man if he can't kill a lion? What if a lion comes? Is my husband going to flee and let the lion eat me and my children or will he fight and save the family?" I think in a lot of societies, there would be many bachelors.

The *morans* go out in groups when they perform this tradition, because this ritual requires the kill to be a male lion. They say a male must kill a male. Before they go out to fight a lion they have a ceremonial dance in which they jump high, straight up and down and chant warrior songs, getting psyched up for the battle. The hunt begins. Lions are brave and don't run from a fight so when he is surrounded, the *morans* chant and beat their shields with their *rungus* (heavy native wooden clubs) or knives, to disorient the lion. Bravery is everything. A Masaai won't run from a fight. They are armed with spears, *rungus,* and homemade swords about two feet long.

The lion studies the warriors surrounding him and spots the one he senses has the most fear. That is the one he charges as he tries to break free. Once the fight begins, the first warrior who puts his spear in the lion is also the one the lion goes after first. When the first spear is thrown, all the Masaai join in on the fight. The Masaai who is the first to put his spear in the lion is considered the bravest and the one who can claim the kill and the mane. He will wear the mane as a headpiece when he takes a bride. The fight can last a long time and many Masaai are wounded and at times even killed during this daring brawl.

If Masaai are guarding the cattle and a lion comes, he must fight or be made an outcast in the community. One of the most wonderful sights is to see a Masaai leaning on his staff, regally standing on one foot, while positioning his other foot on his knee and watching over his herd. A whole pride of lions often come in at the same time, but the warrior must fight them or the lions will scatter all their livestock and have a great feast. The Masaai are the only tribe the lion fears, but when they are hungry and need food, they will come anyway. The lion recognizes the Masaai cloth and their smell. He must be made to know they can't get away with attacking even one cow or there will be a fight.

In the late afternoon, you can see the herdsmen turning their cattle and goats back to the *manyatta* to protect them for the night. At least one *Moran* has to stand guard all night in case the lion comes. It's a hard life. When the lion does come, they are desperate for food and very brave. He will jump the thorny bush fence and pick out an easy prey to grab. Of course, many nights are pitch-black with no moon, but the *Moran* knows a lion is there because of the bellowing and nervousness of the cows. He also hears the roaring and comes to the sound to fight the five hundred pound killing machine, even in the darkness. When the other *moran* in the manyatta hear the commotion or the signal, they come and join him in the fight. The lion will try to take the cow either by going through the fence or jump the fence. The other lions in the pride wait outside the manyatta to join him in the feast surrounding a manyatta at a distance. Then one comes close enough or the wind shifts so the cattle can smell him. When this happens, all the cows could break out of the manyatta in fear running right into the jaws of the pride. This is a constant ordeal that goes on everyday throughout Masaai land.

Most of our Masaai pastors have killed several lions while defending their herds. The tribe honored Loren with a special cloak made of hide, making him an honorary Masaai tribesman. They said this cloak symbolizes being "the bravest warrior," the one who first puts his spear in the lion. They gave this to him for being brave enough to come to them to bring the Gospel. They also honored me with many *shangas* for being a brave warrior to dare come with Loren deep into their land. One day near one of the churches we built, five people were killed and eaten by lions. We had camped right beside this church in tents. Most of our pastors travel the countryside with only a Bible in hand and risk all sorts of hazards taking the Gospel to their neighboring villages. It's amazing that Masaai women will walk through this lion country for many kilometers to come to the church to hear the promises of God. This is where they live and all they know. They have no choice but to be brave.

On another occasion, visitors came from the states to visit the work we were doing among the villages. We arranged for them to stay in a tourist camp for safety. The next day Loren took them to visit some of our churches and left the camp early that morning. I stayed behind that day to do some other necessary preparations. There had been a lot of rains that delayed their return that night. When they had not gotten back to camp by seven, I became alarmed. Loren would always communicate with me somehow if he is out and going to be delayed though we are seldom separated. We had cell phones by now in Africa but the network was down and he couldn't reach me. I knew something had happened.

Knowing they were in the middle of lion country and it was now dark, I found a ranger and set off with him in the direction I thought they might be coming from. By this time it was pitch dark with no moon and you could hardly see the ragged dirt road. It was perilous. I knew they were at least in a four to six hour radius from the camp. The ranger had a difficult time driving because now it had started raining very hard. When we almost ran into a baby elephant, I realized this might not be the wisest thing for me to do. The rain was so hard we couldn't see the baby's mother, but knew she was close by and could ram the vehicle and there would be no way of escaping her.

The ranger had a shortwave radio and communicated with several gates, which were manned by the Wildlife Authority. He asked if any of them had seen our vehicles pass through. Inside the game reserve, the gates close promptly at six P.M. by law. I thought they might be broken down somewhere and was determined to find them, hoping they were already inside the reserve and not locked out.

After about an hour of calling on the radio to the different check points, the driver got a radio call back that someone had seen them pass through his gate. Unfortunately, after they passed through the gate, Loren told me later, one of the vehicles, the Trooper, had a problem. The road was extremely muddy, but the four-wheel drive had broken down. It was totally dark and at this point only drizzling rain. They knew they were in the middle of lion country and this was their prime hunting time. The team had no choice but to get the torches (flashlights) out, and try to fix the vehicle. It was a dangerous situation. There were about ten of them stranded in the vehicles. Incredibly, about two minutes after the breakdown, out of the darkness came several Masaai warriors. At the time, I'm sure they felt like angels and I know the Lord must have sent them with all my prayers and theirs. "Are they not all ministering spirits, sent forth to minister for them who shall be heirs of salvation"? Hebrew 1:14.

They said they would stand guard and protect them while the team worked to get the vehicle up and running. I knew Loren always carried spears and rungus, as a matter of course, in all our vehicles. After about forty-five minutes, they were finally able to move again and actually got back to camp before I did since I blindly went an hour and a half away in the opposite direction, but now headed back hearing they were at least in the reserve. I still had no idea what had happened to them. The camp fed them even though they got in so late. I'm sure their dinner never tasted as good as that night. When I got back we all sat around the table and they told me the story. Right when you think you are alone in a situation, God always sends help. This was our life and we learned to lean and rely on God's help when we didn't know what to do. Loren scolded me for going out in the night looking for them like that, but he would have done the same thing had I been out there. We learned to be each other's best friend, especially in this hostile environment.

A LEOPARD IN THE CAMP

We headed deeper in the Mara down by the Tanzanian border. One of our team members flew with Loren into this remote area. This day the ceilings were low so he also had to hug the earth somewhat. Even though he used the GPS, they couldn't find the airstrip. Suddenly they spotted the old Land Rover in a grassy field with several people standing by it waiving to them. Loren asked, "Are those our people? Where's the airstrip?" Then he spotted where it was. The strip was like a winding farm road with two bare spots where wheels had passed over it sometime before with high grass growing up in the middle. They agreed, "I think that's it." He said he raked up all the nerve he could get and set the plane down weaving back and forth.

We met up with the rest of the team who were staying in an old abandoned building that had open windows and doors. Again, this was lion country and immediately the team secured the building up tight with whatever they could find.

The next day we moved to a village in a jungle area to start construction of another church. The locals assured us there were no lions in this area, nevertheless, since we didn't have a gun, Loren armed himself with a big buck hunting knife and a flare gun. If something bad came, he determined he would not go down easy. Back at the camp that night, we sat around the campfire to eat. The night was uneventful and the first night we had no watchmen. We went to sleep with a fire roaring to ward off any strangers that might accidentally come in. Loren put his spear, buck knife, *rungu* (African club), and flare gun next to our mattress; if something did come, he would fight with what he had. The second night, we felt uneasy and decided to hire two Masaai as night guards for the camp.

About two A.M., we heard a dog yapping loudly nearby and the Masaai began talking loudly to each other. The next morning we asked them, "Why was the dog barking so much last night and you boys were talking so loud?" They told us a big leopard had come into the camp and it had stopped about three feet from the tent right next to ours. That got our undivided attention. Leopards will always attack. Masaai fear them more than lions.

The leopard had come for the dog and the dog, realizing what was about to happen, was smart enough to hunker down between one of the Masaai's legs and barked. All these guys had for arms were spears and knives.

We completed the church and rumbled down that old twisting airstrip harvesting grass the whole way. We zigged and zagged. There was a valley at the end and after using the entire strip before lifting off, Loren put the nose down and flew into the valley to gain airspeed before climbing out and heading back home. He loved flying so much and after being a bush pilot serving the Lord in Africa, no amusement park ride could top this great adventure of real life.

Chapter 13

Vultures

Our next crusade was in Mbarara, a small town of fifty-eight thousand in western Uganda close to the Rwandan border. We flew from Nakuru, Kenya to Entebbe and spent the night. The next day we flew down the western shoreline of Lake Victoria then turned inland for Mbarara. We must have been fifty miles from Mbarara when we ran into a bad thunderstorm, which came up out of nowhere and forced us to turn back to Entebbe. This always unnerved me and is one of the reasons, I'm sure, that I came to be such a prayer warrior. The next day, we flew all the way without an incident.

SOARING WITH VULTURES

We loved Uganda. One of the things about ministry is that we always tried to spend some time in the towns and villages before we ministered. It's a good way for us to get a feel of the place and good for the town to get a feel of us and who we were and why we were here. While our team began setting up the field for the meeting, a pastor friend from Tanzania, came to join us and lead a prayer team for us in this meeting. This pastor had never flown before and wanted to go up with Loren in the plane. He and another African pastor went up for a flight to see Mbarara from the air. They were totally amazed as they looked down to the earth from an eagle's vantage point. Loren always enjoyed taking our African friends up who had never flown before, but while returning to the airstrip, to his horror, he saw a big flock of giant vultures whose wingspans can go up to five feet wide. As they approached each other, he saw terror in their eyes. They met almost eyeball to eyeball. Immediately Loren turned the plane hard to the left and dove like a dive-bomber trying to miss them. Some of them folded their wings like a fighter jet, and also dove straight toward the earth. They all tried to miss each other. The birds were so big, for sure if they had collided, Loren would have crashed. Loren said the African friends turned as white as ghosts. I'm sure that night they were the talk of the vultures on their roost. They must have said to each other, "Man, what kind of monster bird was that?" Our African friends were so petrified with fear that none of them wanted to talk about it. They were literally scared speechless. After that incident, I didn't want Loren pulling the banner in this area. We noticed the sky was full of vultures and the winds were unpredictable in this area. I asked him not to fly again until it was time for us to leave. He agreed.

Mbarara became well-known internationally because of a cult mass murder that occurred in that city in the late 1990's. There was a man in Mbarara who called himself a prophet of God. He worked with a former nun who claimed to be a prophetess herself. They declared Jesus was coming on a certain date and told the people of their "church" to sell all their possessions and bring the money to them so they could warn other people. The people were told after Jesus came on this date they wouldn't need money or possessions anymore. On the date of His supposed return, over one thousand people gathered inside their "church" building waiting for the Lord to come and take them away to heaven. When everyone was inside, these false prophets locked the people in and set the building on fire. Everyone inside was incinerated, a horrific mass murder. The so-called "prophet" and "prophetess" could not be found. They had fled with the people's money. In the wake of this terrible history of Christianity in Mbarara, the tragedy left a bitter taste for the Gospel in this area.

THE MBARARA, UGANDA CRUSADE

The first day of the crusade, we had around five thousand in attendance. The people in town had gotten to see our daily lives for a time and watched as the men had worked on the field in town. Nothing could be hidden and the people were happy to have us and felt God would indeed be in this meeting. So many people were born again and gave their hearts to Jesus. Many great miracles of healing happened. One man testified thugs had attacked him and his nine companions. All of them were killed except him, but he had been shot badly. God miraculously healed him of his injuries. Another testified of being healed of a stroke, and another of paralyzed legs. It was amazing to be a part of the restoration the Lord did in this place.

One thing that became a daily prayer was the weather. It had changed since we arrived and every day when it was time for the crusade to begin, heavy black storm clouds moved in. We prayed in the name of Jesus the Lord would hold the storms back. The clouds would stop within a half a kilometer of the crusade. This happened every day and those storms never affected the crusade. It was truly a sign and a wonder. We were later told Muslims had slaughtered a bull as a sacrifice to put a curse on the crusade. They said they prayed the rains would come and stop people from coming to hear the Gospel, but it didn't work. The last day virtually most of the whole town came to the meeting, and there was a breakthrough of the Holy Ghost. The Apostle Paul asked, "Have you received the Holy Ghost since you believed? Acts 19:2, 4 and 6. This makes clear that it is a definite separate infilling after salvation. Jesus Himself said in John 20:22, as He was speaking to the disciples after the resurrection and before His ascension," receive ye the Holy Ghost". Jesus has not left us alone, God is still with us and He wants to comfort us with His peace. After such a terrible tragedy having happened in Mbarara in the name of Christianity, the people's faith had been restored in the Lord and the city was healed.

KAMPALA, A STRANGE PRAYER MEETING

After closing out the crusade in Mbarara, the whole team headed for Jinja, to the east for another crusade. They took the truck and we flew in the plane. We had to pass through Kampala on the way and knew the pastor of a large church there and decided to stop and just say hello. We had met him a number of years earlier and preached a conference and a revival at his church in the past. We noticed he had become popular in America and was on television regularly, but we had not heard him preach since we were there about eight years before. We were not current on what was happening with his ministry. We landed in Entebbe and drove to Kampala to his church, but he wasn't there. The under pastors knew us and gave us a warm greeting. They said on Friday night the church was hosting an all- night prayer meeting in the big city stadium. They asked if they could rent our crusade sound system and also come to the meeting ourselves since we weren't due in Jinja for another week. We told them our equipment was not for rent, but because they did God's work, they could use it with no charge. We instructed our team to take the equipment there, set it up, and operate it.

Friday night came and we met the pastor at the stadium. We embraced each other and chatted. He asked if Loren would be one of the preachers that night and he agreed. About fifty thousand people showed up and in the early part of the meeting the praise and worship was good, with different pastors leading the people in prayer. We were seated on the platform with other special guests.

After the meeting had gone on awhile, things became bizarre. Following wonderful prayers offered by different pastors, there would be different music and dance groups. The problem was they sang and danced exactly like the world. One young woman in a singing group had on a tight mini-skirt and she shook her backside. This was supposed to be a prayer meeting. Soon it felt more like we were at a Michael Jackson concert. They would pray awhile, and then rock awhile. This went on all night, back and forth. We became increasingly uncomfortable. When the minister got up to take an offering we got annoyed because he hustled the people. He told them, "If you sow what the servant of God tells you tonight; If you came in a *matatu* [a public transport minibus], this time next year you will be driving your own car. If you are renting a house, after you give this miracle offering, this time next year you will own your own house." This continued until we looked at each other, reading each other's mind, "How fast can we get out of this place?" God was not within a thousand miles of this so called "prayer" meeting. Our spirits were grieved.

There was a famous Gospel musician from America who spoke up and publicly gave five thousand dollars in the offering to support this ministry. The people couldn't run to the altar fast enough to give after that. Then the pastor began to cast out devils. We know it is scriptural to cast out devils, but what he did was nothing but a show. He brought up a young woman to the platform who manifested demon possession, writhing, twisting and crying out. Instead of casting the devil out of her, he let her demonically manifest for about an hour in front of this whole multitude. It didn't glorify the Lord. It glorified Satan. The crowd was mesmerized, but it wasn't by the power of God.

BLASPHEMY

The pastor proceeded to tell the crowd that at other outdoor crusades he had held in other countries, at times a white dove would show up at his services. He planted that thought in the crowd's minds several times throughout the night. At about four A.M., we heard the multitude start shouting in incredible excitement and looking up. It was a white bird flying around the inside of the stadium. The crowd went wild and was in a frenzy. You could tell the people thought this was a dove, the Holy Spirit personified. The problem was, it had a long neck and wide wings. It looked more like an East Texas cowbird to us. For sure we knew it wasn't a dove or The Holy Spirit. We watched as that bird flew around and the crowd screamed louder and louder. We thought, "Oh God, if that bird comes and lands on that pastor's shoulder, they will all think he is Jesus Christ, the chosen one.

We looked around at the hundreds of other pastors on the enormous platform who were totally mesmerized. There were other visitors from the states and they were all in a state of awe. By now we were sure God was not within a million miles of this place. The carrying on there was blasphemous. We got up and walked off the platform and out of the stadium. It was 4:30 in the morning and we returned to our hotel tired and somewhat trying to make sense of what we had witnessed. We fell asleep but about six A.M., the Holy Spirit woke Loren up and these words came to him, "A strong delusion, a strong delusion. "... because they received not the love of the truth, that they might be saved. And for this cause God shall send them strong delusion, that they should believe a lie:" 2 Thess 2:11". He woke me up and told me what the Lord had impressed in his spirit. "O Lord," we cried, "What has happened to your church, your bride? "...even as Christ is the head of the church..."Ephesians 5:23. Lord God, help us."

We tried to figure this thing out and knew birds can be trained. The other explanation was that it was witchcraft. This was a literal manifestation of what we had been warning Pastors about in the conferences we had been doing trying to bring them back to Biblical Christianity. It had been demonstrated before our eyes in an astonishing, reckless, and shameless manner. We knew there was no turning back for us now regarding speaking out and standing up for sound doctrine. Matthew 24:4 warns us to "take heed that no man deceive you." This was actually a traumatic experience for us and most of the multitudes of believers were completely deceived.

RAFTING THE NILE

We journeyed on to Jinja. This was a fascinating and beautiful place. The first two weeks we weren't aware of where we actually were. We began working hard to prepare for the crusade, which included meetings with the local pastors joining together for the salvation of their city. We were ready and anxious to begin the crusade. Our advance people had come in earlier and these last two weeks we finalized the preparations. It usually takes nine months to a year of advance work to prepare for a big crusade. Our team was also still stunned at what we had all witnessed in Kampala. It made us want to do the real work of God even more than ever. Scam meetings like the one we had witnessed makes evangelization even harder.

The city of Jinja is situated on the Nile River and at first glance, in the city, the river flows in relative calm. This is the spot so many explorers trekked throughout Africa searching for, the source of the Nile. Lake Victoria pours a phenomenal amount of water into the Nile and this was the origin point. There was a monument to Mahatma Gandhi of India in a beautiful garden and his ashes were said to have been scattered at that point in the Nile.

Massive underground rivers billow up at this point. The incredible amounts of water coming from Lake Victoria and the underground rivers have been flowing from the beginning of creation; a phenomena and the miraculous handiwork of God. These waters of the White Nile flow from this point in Uganda and join the Blue Nile up through Sudan and continue up through Egypt. It's the only major river that flows from south to north.

Two days before the crusade started, the pastor's committee took us out to relax and to show us around the city and the outlying villages. We had no idea what a "wonder" lay ahead of us. They drove us downstream a little way to see some of the villages and a different view of the Nile. We were shocked when we saw the mighty rapids as the river literally thundered north. A company from New Zealand had recently set up a rafting business and there were tents scattered around where people from all over the world came to camp and ride the rapids. We did not know that this existed and was a world-class rafting destination. We sat down to have a soda and watch one of the great wonders of God. It was breathtaking. We actually saw some of the smaller rapids where the guides schooled the rafters. They taught them what the commands meant and what to do if you fell out of the raft and into the river. From our vantage point we could see only two sets of rapids and people rafting them. It looked like a lot of fun and we decided to go rafting the next day and bring the team with us. This would be a once in a lifetime experience for us, and since the team never gets a chance to do anything like this, we wanted to treat them because they had worked so hard. We all needed some R&R before the crusade began. We went back to camp and told them what we had planned. They were excited, not knowing what to expect.

The next day we and the team met with our guide who would navigate us as we rafted the river. Here is the picture: eight men and one woman, seven African men who had never lived around water and did not know how to swim. None of us had ever rafted in our lives. First of all, we were given lifejackets and helmets. Our men had never worn anything like this in their life. Loren and I put our gear on while the men walked down to the river where our raft was located. When they got within view and actually saw the rapids they suddenly threw on the brakes and stopped in their tracks. I will never forget the face of one of them as he stared at the churning water. He turned and then looked back at us. You could tell he and the rest of the team did not want to go. It no longer looked like fun to them. Then they fixed their eyes on me putting on my life jacket. Later they told us, "None of us wanted to go, but if Mama went, we had no choice."

We all got into the large raft with our guide. He showed us how to "fall into the water" and assured us that the rapids would "spit us out" after a few seconds and a kayaker would be right behind us to pick up anyone who got dumped. He showed us how and where to grab the kayak if this happened. In faith I had already assured the team that not one of us would fall out of the raft. We all had prayer together. Still, it was a bit intense and we had only seen the smaller rapids. Now it was time to do it. After telling us about the commands and showing us how to obey his commands, the guide pushed us off and we began to row. No one could slack off; it was a team effort. We would go over the rapids together or be upended together.

We had agreed to go for the shorter trip that went twenty-five kilometers because we only had this one day to do it. There was a two-day trip available but we had to preach the next day. We novices began to row, following the instructions barked out by our guide. Our adrenaline pumped as well as an underlying fear of the unknown for what might be ahead of us. We made the first set of rapids and they weren't as bad as some feared. Then came the second set, which were quite a bit rougher. By the third set, we all had a ball and felt that we could do anything. We came to calm water and were told there were no crocodiles here, so Loren jumped in the water to cool off, and slowly floated downstream alongside the raft. It was wonderful enjoying the beauty of nature along the shores of the Nile, and to see so many species of wild birds flying around and diving into the water. Even in areas where the guide said there were crocodiles, we saw children with their naked little behinds diving and splashing with cheerful abandon. Oh, the joys and dangers of innocent childhood. Then we approached the more difficult rapids. We soon began to realize the ones we had seen before were warm-ups.

The rapids were rated by difficulty, from one to six, with six being the most dangerous. We went up to category five rapids for the rest of the way. The exhilaration was incredible. The thrill was a maximum ten. We had to hang on for dear life, nearly being thrown out of the raft going over each set. Some rafts ahead of us had totally flipped upside down unloading all their passengers. At one point while we approached a set of rapids, our raft was turned around by the churning force of the water, and horror of horrors, we went over the falls backward. We were told if we were thrown out and covered by the torrent of water, "Don't panic, count to eight. By the time you got to eight, you would pop up." I don't know where they came up with that scientific formula. The problem is being underwater and tossed around like you were in a washing machine could make you stutter or forget to count. We stood by our prayer.

About halfway through the twenty-five kilometer trip we approached a small island in a calm area and went ashore. The rafting outfit had a crew there to meet us with hot coffee, tea and sandwiches. We were freezing cold from the water and they had a big roaring open fire to help us warm up. It was a nice break from the intensity of the morning run, but we were all talkative about our experiences. After a short break, we continued the trip, riding the rapids like cowboys riding wild broncos in a rodeo. About this time, the guide told us that the upcoming rapid was a six, difficult and the maximum on this leg. Anything above a six was certain death. He gave anyone who didn't want to risk it the opportunity to get off and walk around it and meet us on the other side. Two of our men opted to get out, but Loren and I with the rest of the team decided to ride it out. Later, we realized one thing we didn't reckon on; losing the weight of their two bodies, which meant that the raft was less stable. This last set of rapids shook us violently. It was so rough, it turned us every which way but loose, but we survived with a story to tell. Now, we approached a rapid that was over a six, which meant it was suicide, and of course we didn't want to throw the dice on this one. We headed for the shore; everyone got out and carried our raft on a trail around this one. The last one for the day was another six and we got back in and rode it out, feeling victorious, conquering our own fears. At the end of the twenty five kilometers, the rafting company had barbequed goat waiting for us and we were congratulated and told stories about our bravery. The team was glad they went on this trip, but none of them wanted to do it again. We got the T-shirts, "I rafted the Nile".

There was a bus waiting for us and on the drive back to our starting point, everyone was lost in their own silence reveling in the experience we just had. We stared silently out the window as the bus passed the local people who lived along the Nile. The children would wave to us, enjoying the evening out in front of their huts. They loved to watch the crazy *wazungu* pass them by, but mostly the team was stunned at what they had just done and at their own bravery. Our minds were filled with the stories about some of the earlier adventurers who went down these rapids before they were charted and died, crashing into rocks and drowning. The peace of our heavenly father enveloped us as we turned in our gear and then sobered up in the reality of what could have happened. We were far south on the Nile, a great distance from Egypt where Moses had been put in its waters when he was a baby. This same Nile was the one that God, through the hand of Moses, had turned into blood to show Pharaoh God's awesome power. Incredibly, these great waters of the Nile had been flowing ceaselessly, from its great fountains of the deep since the creation of the world. It inspired us anew about the mighty wonders of the power of God through His creation. We thanked Him again for the opportunity to live another day to serve Him.

THE JINJA CRUSADE

Once again Loren was able to advertise by flying the banner behind the plane for the upcoming crusade. This time, thank God, it was uneventful. Jinja was primarily an Islamic city, but God showed Himself strong again. The crusade was carried live on the radio. A crippled woman from a distant village heard the message of Christ's saving and healing power and right in her home she found salvation in the Lord and received healing. She came out to the meeting and testified how she had placed her trust in what the Lord had done for her on the cross at Calvary. Her soul was safe in the love of God. Romans 10:9-10. "… if thou shalt confess with thy mouth the Lord Jesus, and shalt believe in thine heart that God hath raised him from the dead, thou shalt be saved. For with the heart man believeth unto righteousness; and with the mouth confession is made unto salvation". God's great plan of salvation was at work again as multitudes gave their heart and lives to the Lord Jesus Christ.

A BIG MARLIN

We went home to the states for three months to be with our family and raise funds for our return trip to Africa. During this time at home, we went up to the northeast to minister in a pastor friend's church. While with them, they told us they had a time-share apartment at a resort in Cabo San Lucas, Mexico and wanted us to use it for a vacation as a love gift. That was such a kind gesture and we needed the rest.

Ever since Loren was a boy he had dreamed of catching a marlin. He didn't know where that dream came from because he had never known anyone who had caught one. In his opinion they were one of the most beautiful and magnificent creatures God ever created, but growing up in a pastor's home of modest means, going after a marlin was a big reach for him. He did not want to spend a lot of money to do this, but the desire continued to linger in his heart and now the opportunity might be available on this trip.

His eyes got big when he heard their offer of a vacation in Cabo. He knew this was one of the best places in the world for Marlin fishing. It was then that we determined to go after one if we could possibly afford it. My daughter worked for the airlines and she offered us free passes to fly to Cabo, so the cost of the flight was taken care of.

We arrived in Cabo excited with the blessings we had received and looking forward to a week of rest and sun. At the airport we were inundated by agents selling time-shares for vacation apartments. We didn't know Cabo was a hotbed for real estate. We had lived in the bush of Africa so long we had no idea of what went on materially in the outside world. There were droves of agents meeting everyone getting off a plane trying to entice them to "come and have a look." One of them came up to us offering great perks to take a couple of hours to look at their time-share apartments. The perk that got our undivided attention was when they offered us a Marlin fishing trip. We were honest and said we were not interested in buying a time share, but they insisted if we would just come and take a look, they would give us a marlin fishing trip in our own boat for only two hundred dollars. Instantly we told them, we'd go.

Our next crusade was to be in the Congo and we had been praying for those meetings, believing God that it would be the biggest harvest of souls we had ever had. At the same time, Loren was convinced God would bless him with a big marlin on this trip. His faith was strong in both.

Although I was not big on going out to sea, I wanted to support Loren and went with him. I agreed with him in faith that he would get his marlin that day. We got up early and reached the docks before dawn. We were delighted to see it was a nice boat manned with a professional crew. As we left the beautiful harbor in Cabo we were full of excitement and expectation. We were barely out of the harbor when we started trolling. Loren was shocked and happy to start trolling so close to land and we were thrilled and felt blessed to see the dolphin racing alongside our boat as we headed further out to sea. The sun was shining and it looked like it would be a great day.

In about an hour and a half he hooked up with a four foot bull Dorado. What a fight. We saw a tall fin of a big shark and the captain asked if we wanted to go after him. Without hesitation Loren told him, "No, I want a marlin." After being out about three hours we were still trolling only about three miles from shore; the captain said the water was very deep there. The hand on deck pointed behind the boat and shouted, "Marlin." I grabbed my camera and got ready for action.

You could see his bill, dorsal, and tail fins sticking out of the water as he basked in the sun. Loren's blood pressure must have instantly shot up he was so excited. The captain turned the boat around and we made a pass close to him. The deckhand reeled in the teaser baits and cast bait in the water. As soon as the marlin saw it, he exploded and tore out after it like he hadn't eaten in a month. He hit the bait with his big bill to kill it. This loosened the line that was connected to the outrigger. Then he took it. As soon as he bit down good, the deckhand jerked hard, trying to set the hook. A marlin's mouth is as tough as a two by four. At the same time the captain hit full power on the boat to help set the hook.

As soon as he was hooked he began to leap out of the water like he tried to fly. The deckhand told Loren to sit in the fighting chair and handed him the rod and reel. It was on. The power that marlin had was phenomenal as he leaped, dove, and ran. The reel sounded like a siren the line went off so fast. After a while he slowed down. I could tell it was a real task turning his head. Loren put his feet up on the railing of the boat as he sat in the chair trying to horse him in. After a long tough fight, we could finally see color as he reeled him up close to the boat. It was absolutely exhilarating for all of us. I could see Loren was ready for the battle to be over because he was tired, but the fish had other ideas. When he saw the boat, he instantly turned and took off harder and ran longer than he did the first round. Loren said his back and arms ached, but he wasn't about to let anyone else have the pole. He finally stopped running the second time and now it was a mighty tug of war. With both hands Loren pulled up as hard as he could, which was difficult, but when he reeled, he could get a half a turn on the reel. Eventually he wore him down, but it was a close call as to who would give up first. He was a good-sized striped marlin. The captain told Loren he was the luckiest guy in the world to catch a marlin in such a short trip.

It was a fantastic day and we had a wonderful celebration. I broiled some Dorado for dinner in our apartment and we had a delicious meal that evening and thanked the Lord for giving Loren a desire of his heart. We felt that this was also a sign that Congo would be the greatest crusades we had ever seen.

Chapter 14
Cannibals

W e prepared to take the plane and fly into the DRC, The
Democratic Republic of Congo. This three thousand mile
flight could be a treacherous journey and Loren did his home-
work making sure the flight plan was thorough and accurate.
We had to get clearances to fly from Kenya, Zambia, and the
Congo, countries we would be flying through to our destination.

We had sent our team ahead of us with all our crusade equipment. Loren and I waved goodbye to the Pokot and took off
from the Chariot of Fire, Pokot International Airport. They were
proud of the airstrip they had carved out of this harsh land for
us and are the ones who gave it that name.

We flew south past Nairobi and headed southeast over Tsavo
where the film the "Ghost and the Darkness" told the story
of building the railroad between Nairobi and Mombasa and
the man-eating lions that attacked the workers. We continued
eastward toward the Indian Ocean and then due south toward
Dar es Salaam, Tanzania. We breezed through the necessary
checkpoints in Dar .We both wore pilot uniforms and found it
was a lot easier to cross borders being clearly identifiable, even
though we were in a small private plane. Our ID tags hanging
down from our necks were easily readable and our photographs
clearly identified us. When they asked me what I did, I never

blinked an eye. I said, "I am the navigator." We have had fun about that one. It was true though because Loren taught me to handle the GPS and read the maps. We worked as a team in everything although I never had a desire to be a pilot. We refueled in Dar and headed south toward Mbeya, Tanzania. This was a long stretch over a vast wilderness full of every kind of wild beast. We had noted dirt airstrips off the WAC charts and stored them in the GPS and noted unmarked ones along the route in case we had any problems or ran into foul weather. Flying in Africa, you must always have a Plan B, C, and D and be ready to deploy in a moment's notice.

We landed in Mbeya and requested fuel, but they had little and it was expensive. We decided to wait until our truck caught up with us. They carried our drums of AV gas. Out in the interior you cannot count on anything, so you must prepare for the worst-case scenario all the time. We went to the local guesthouse to get some rest while waiting for our truck to arrive. During this waiting period, Loren came down with malaria, but thank God it wasn't too bad and the Lord touched him and he soon recovered. When our truck arrived two days later, we met them at the airstrip and discovered a lot of fuel was missing in our plane. Loren always knew where he was on fuel, flying long distances over the jungles. It became obvious someone had raided our tanks. Then, out of nowhere some men showed up trying to sell us some aviation gas at an expensive price. We became suspicious because the amount they had for sale was the exact amount missing from our tanks. As a pilot, you never trust your gauges, but always check the fuel manually. He had done this before leaving the plane when we landed and knew exactly what he had. It was upsetting. We informed the local police, but they wouldn't do anything about it. It was a stacked deck.

After refueling from our own drum off the truck, we took off and headed toward Zambia. We climbed to ten thousand feet to make sure we had good clearance and visibility because the WAC charts we used were not detailed. Even though it wasn't cloudy, there was a lot of clear-cut burning going on. The people set the forest on fire to clear off the land so they can farm, but it is environmentally disastrous. The air was filled with smoke and the visibility was bad; it was nearly IFR. This was our first flight into Zambia and Congo and we were nervous flying over this desolate territory and into unpredictable airspace.

Our route took us over a vast swamp land in Zambia. Loren didn't tell me, but he was sure praying the plane wouldn't get into any mechanical problems because there would be no place for us to put her down. He knew those swamps were full of crocs, hippos, and snakes. He told me to look down and enjoy the beauty, but I refused. I had my eyes fixed on the Bible and the ninety-first Psalm. It had become my favorite book and chapter.

We had planned to land in Ndola Zambia and about a hundred miles out picked up its signal and tracked the beacon in. However, not long after picking up the Ndola's signal, it went off. Now the only thing we had left to navigate with was the inexpensive GPS. It pointed in the right direction, but we were apprehensive because it was so smoky and we were getting low on fuel. There was no turning back. The GPS gave us a countdown of mileage but still at eight miles out, we could see no city. Finally when we were almost on top of it, it came into view. What a relief. We re-fueled and spent the night.

The next day was the last leg to our destination city, Lubumbashi, in the southeast part of the Congo. Although we had clearance to fly into the Congo, it was an eerie feeling because we knew they had been in war for some time and proceeded cautiously. As we entered into Congolese airspace, we tried to contact approach control, but never got a response. That was not a good feeling, but we had no choice but to continue toward our destination. As we approached Lubumbashi, he tried contacting the control tower again, but still there was no answer. Fuel was now an issue; we had no choice but to land.

Setting up for the final approach, I said, "Loren, look—a big anti-aircraft gun is tracking us."

He told me, "It's too late now; I have no choice but to land. We can't go back." Thank God they didn't shoot but that big gun tracked us all the way down. Right off the runway was the wreckage of an airliner that hadn't made it and had been shoved off out of the way. This was our "welcome" to Congo.

Heavily armed soldiers were all over the airport, but thankfully our crusade manager was already there with some bishops and host pastors and gave us a VIP welcome. We were greatly relieved and thanked the Lord this flight was over and we had arrived safely without incident. We were covered in sweat, not only from the heat, but also from the stress of the last few days of flying over the jungles into the unknown. Our welcoming delegation took us to our hotel under heavy guard. In fact, two armed soldiers were assigned to guard us day and night for the duration of our stay.

The Congolese government had also granted us a permit to pull a banner to advertise the crusade. A few days later after some rest and meeting our ground team of bishops and pastors, Loren started his preparations for that event. The banner was much longer than usual because it had to be written in the local French language. After setting everything up in the usual way, he took off and came back low, flying over the grassy area, where the banner lay on the ground. With the focus of a fighter pilot, he swooped down and snagged the rope with the grappling hook. He pushed the throttle all the way in and pulled up, but unbelievably the plane wouldn't climb. That was when he realized the banner was too long and he had mistakenly calculated the length of it for sea level but Lubumbashi's elevation was four thousand feet. The air was too thin to pull that much. He jettisoned the banner before he got to the end of the long runway; otherwise he would have been pulled to the ground.

ARRESTED

The next day we shortened the banner and he prepared to go up again. As he got ready to get in the plane, a man came up to him in plain clothes wanting to ride. Loren told him it was against aviation regulations for anyone but the pilot to be in the plane while pulling a banner, and gently refused him. He didn't like it, but he wasn't about to break aviation regulations and got in and took off.

This time he snagged the shortened banner successfully and flew around the city. Pulling a banner behind a plane is a tricky maneuver. He had to fly about seventy-five miles per hour with ten-degree flaps to help maintain lift. He scanned his gauges and to his chagrin saw the temperature gauge in the red. If this continued, it would cause engine failure. Realizing the crisis, he was about to jettison the banner in an open field and speed up to help cool the engine down, but before doing this, in a last ditch effort, he pushed the fed fuel knob in to full rich. To his relief, immediately the temperature gauge went into the green and he was able to continue with his plan. It was so interesting from his vantage point in the air, seeing the people on the ground coming out from everywhere, looking up. They had never seen a plane pulling a banner before and our hope was it would interest them enough to draw them to the meeting and look up to Jesus. "... look up, and lift up your heads; for your redemption draweth nigh". Luke 21:28.

Returning to the airport, he jettisoned the banner next to the runway and felt quite good because he had informed thousands of people about the upcoming crusade. As he got out of the plane, the authorities met him and Loren was arrested and taken to the control tower. We were confused. The man arresting him said he never should have made the flight with the banner without one of their men riding in the plane with him. Loren answered them boldly that it was against aviation law, worldwide, to have someone in the plane while pulling a banner and he did not want to break the law. We had spent a lot of money to get the permit to pull the banner. The proper authorities in Kinshasa had authorized it and all his paperwork was in order. This began a big argument, but Loren stood his ground with the security officer. He told him he didn't even know who that man was as he had not identified himself in any way. Even at that, aviation law forbids him to have a passenger riding with him while doing this procedure. The chairman of the crusade and other influential men hosting us were called to the airport to help sort this dilemma. What a fix; arrested in the Congo and in an unpredictable situation. After key bishops negotiated with the authorities for several hours, they finally gave in and released him. It's never dull working for the Lord. As we thought on it later, because of the war, I'm sure there may have been "security areas" they wanted to be sure he did not see from the air. God protected us from a lot.

LUMBUMBASHI, CONGO CRUSADE

After an unexpectedly long and difficult trip, our truck and team finally got to Lubumbashi just two days before the crusade was scheduled to begin. Normally our ground team arrives in a city at least several weeks to a month ahead of a meeting because it usually takes that long to set up. Now the team had no choice but to try to get two weeks' work done in two days. Our permits defined the time limitations on us. Even after hiring extra help, the team was still setting up equipment at the last minute before the meeting was to start. They had to work around the clock with almost no sleep. Unknown to us at the time, this field was the main ground where local witches used to work their magic. We later discovered there had never been a successful meeting on this field.

For our own safety, the authorities had given us heavy military escort to the grounds. We arrived with our vehicle everyday surrounded by foot soldiers. The power of the Gospel shook the people. The war in the Congo had been going on for so long that people saw their need for God and the salvation promised through His son Jesus Christ. "Neither is there salvation in any other: for there is none other name under heaven given among men whereby we must be saved". Acts 4:12.

One night a seven-year-old boy who had never walked came walking across the platform testifying Jesus had healed him. The multitude was electrified. A little girl about six years old told us she had been totally blind. She said the Lord had given her sight. She could see perfectly and reached out and touched Loren's nose to demonstrate that she could see his nose. The Bible says, "and they brought young children to Him, that he should touch them; and his disciples rebuked those that brought them. But when Jesus saw it, he was much displeased, and said unto them, Suffer (allow) the little children to come unto me, and forbid them not: for of such is the kingdom of God. Verily I say unto you, whosoever shall not receive the kingdom of God as a little child, he shall not enter therein. And he took them up in his arms, put his hands on them and blessed them". Mark 10:13-16.

People brought their witchcraft paraphernalia and burned them in a big drum while the crowd cheered, danced, and sang with joy. The city was being set free by the preaching of the Gospel. The coming of Christ into this world was for the purpose of giving His life, a ransom, thru his death, burial and resurrection. God knew where they were and had sent His Son to save their souls. The "whosoever" of John 3:16 included them.

CANNIBALS

Some young people sent word thru our hosts that they wanted to talk to us privately. I went over early the next day with some guards to meet with them before the official meeting began. There were three children; a seven-year-old boy, a nine-year-old girl, and their thirteen-year-old sister. Their story was incredible. The children said their mother had been a witch and ever since they were very little had taught them witchcraft. The boy said he could turn himself into an animal or a bird and could even fly. He said he could turn himself into a lion and kill and eat people. (These accounts are common in the deep interior of Africa.) He told me and the team with me that he could levitate

253

and shared many other astonishing stories. The children continued telling me that on one particular occasion, it came time for them to make a sacrifice to Satan so they informed their mother that she would be the sacrifice. They confessed to me they killed and ate their mother. Now it was time for another sacrifice and they had put a curse on their father to kill and eat him. In the middle of all this our crusade came to the city. Naturally curious, the children came with the crowds to see what was going on in this big meeting. Once at the crusade and hearing about how Jesus was God and came to earth to be our sacrifice, they became convicted of their sin, of what they had done, and what they would do. They told me they wanted to get right with God and wanted us to pray the curse they put on their father would be nullified. I took time and explained the gospel carefully again to them; how Christ had come to reverse the curse that sin had brought to us all, the law of sin and death. "For the law of the Spirit of life in Christ Jesus hath made me free from the law of sin and death". Romans 8:2.

I prayed with them and they verbally repented of their sin. They were marvelously saved. I also prayed for the father to recover from this curse and also come to the Lord thru this meeting. The children came to the platform that night and told the crowd their story and that the power of Christ was greater than the power of Satan.

We discovered cannibalism is still very prevalent in Congo along with trading in human skins. The newspapers carried such stories all the time. Thank God for Jesus Christ and His marvelous saving grace and forgiveness. His ability to change lives is undeniable.

The last day of the crusade as we entered the field, our group of soldiers surrounding the vehicle, we had never seen such a crowd. You could not see the end of the multitude and it appeared everyone responded in a great roar of confession to repent of their sins and accept Jesus Christ as the Lord. It was worth everything we had gone through to get there.

We had made a strong friendship with our host pastor who had a church of more than seven thousand in the city. Our field workers and counselors came from his fellowship and other supporting churches in the area. They continued as a team in following up the vast number of new converts to Christ. This Pastor's quiet strength and great faith was a special blessing to us and we worked well together.

As we got ready to fly out, we asked a fellow pilot at the airport if he knew of some good alternate airstrips we could use if we ran into bad weather or had engine problems. He answered calmly and said, "Yes, but I wouldn't recommend landing anywhere if you can help it because there are too many cannibals out there."

THE HUGE LIKASI, CONGO CRUSADE

No one has ever called us cowards. We had great expectation for this city called Likasi. As we traveled on the road now, deeper into Congo, we noticed people selling gasoline along the roadside in five-gallon jerry cans. We were advised never to buy from them because these entrepreneurs mixed tomato juice and water with the petrol to increase their profits. When we got to Likasi the looks of some of their vehicles entertained us. It seems no car ever retires in the Congo. When it breaks, they cannibalize other cars, even body parts, so a car could be made up of many different models and brands. It reminded us of Johnny Cash's song, "I got it one part at a time and it didn't cost me a dime." They were a sight to behold and it got more than one grin out of us.

Our advance man gave glowing reports of the great reception we were going to receive in Likasi. He predicted we would have a million people turn out for the crusade. After we arrived though, it wasn't what he said; things had fallen apart. We discovered we actually had only one pastor cooperating with us. We had expected the biggest harvest of souls we had ever seen.

Furthermore, the population was not a million as he had told us, but only six hundred thousand because of the war.

The Belgians had colonized Congo in the past, but they were kicked out when the Congolese fought for their independence. Now the roads and all public services were in total disarray. Many of the buildings in Likasi had been blown up in recent fighting and you could see big caliber bullet holes on many walls. Our hearts went out to the people for all they had gone through, and we realized it was a dangerous place.

We were put up in a small apartment with four heavily armed soldiers guarding us 24/7. This environment made our stay depressing. We were virtual house prisoners and couldn't go out anywhere. We had instructed our crusade manager to get a massive field because our expectations were so high based on the feedback we had gotten from him. The field he found would hold a million people standing, but now that we only had one pastor helping us and the population was six hundred thousand, we knew this could be a humiliating experience and on the surface could be a disaster.

The first night God stirred the people in such a way that all the other pastors in the city heard about it and decided to join us the next night. Working for God is such a step of faith in everything you do. The problem was the field was *so big*, even when the crowd reached two hundred thousand it still looked like a few people having a Sunday afternoon picnic. Loren fought depression. Every night after ministering, he would go under the platform, which also served as our private waiting area and sit with his head down holding it in his hands. When your expectations are so high and what you see seems so low, it is a real disappointment. I tried to comfort him by emphasizing what God had done with those who were there, but he kept slipping lower and lower as the week went by. We knew this was his flesh reacting, but he couldn't seem to get a handle on it and he realized this was a spiritual battle and had to conquer his feelings. "..For they that are after the flesh do mind the things of the flesh; but they that are after the Spirit the things of the Spirit. For to be carnally minded is death; but to be spiritually minded is life and peace". Romans 8:5-6. We spent the mornings in prayer and preparing to minister. We later found out our crusade field was the place where witches did their pagan rituals. We were told many people were *afraid* to come to that field because of its history.

The Lord is so marvelous. One night as Loren preached, a young boy about eight years old who was born deaf and dumb was miraculously healed. He could hear and he began to speak. It was a marvel and faith began to build. By Saturday night, we had thousands in attendance, but again, the field was so massive it looked like a big picnic. Although Loren's spirits were down, like Abraham, we hoped against hope that tonight God would supernaturally move the whole city. As we drove to the grounds, we looked around and were astonished at what we saw. Literally rivers of people were headed for the crusade from every direction of the city. We marveled at the great sea of people. We had never witnessed anything like this in our lives.

The entire multitude seemed to respond in unison when given the invitation to accept Christ as their Savior. The work of the Holy Spirit is something we are continually amazed at. One pastor testified he had over one thousand new people come to his church the Sunday morning before our last service. He had to tell his first service to leave to make room for the new people and a second service.

MBUJI MAYI/ Diamonds

From Likasi we had to go back to Lubumbashi and re assess how to get to Mbuji Mayi. It was a nine hundred kilometer trip from Lubumbashi to Mbuji-Mayi and the roads were virtually impassable. The only alternative was to put our truck and generator on a train and send it by rail the rest of the way to Mbuji-Mayi. The wheels of the truck had to be taken off so it could be set on a railroad car. Otherwise, it would be too tall to pass some areas. The truck and trailer were loaded onto a rail car, but then had to wait for a locomotive to come and pick it up. The team was told the locomotive needed to be repaired first. It delayed and delayed and delayed for over three weeks.

Our crusade manager was already at work in Mbuji-Mayi putting the crusade together; but seeing the problem with the train forced him to fly back to Lubumbashi to try to get our equipment moving. All of our advertising: the posters, hand-bills, platform, lighting, generator and P.A. equipment were all on the truck. We needed to start advertising to mobilize the city. Things were much more costly in Congo and the team had to be careful on their spending so our Pastor friend was kind enough to put the team up again at his church while they waited for the train issue to be resolved. Time was money and money in this case was double and triple the estimated original budget to keep our equipment moving and our team fed and housed.

We were assured the locomotive that was to take our truck and trailer to Mbuji Mayi was repaired and almost ready to go. Thinking they actually meant what they said, Loren and I decided to fly on into Mbuji-Mayi and meet the host pastors and bishops and to get acquainted with the city. This time we took a Congolese airline; an old DC-6 prop passenger plane, which had turbine engines mounted on it. The DC-6 was so old it was like a time machine. Some of the seats didn't have seat belts, only a rope to tie yourself down in. We had never seen so much stuff and people crammed into a plane in our lives. It was evident the pilot did not look at a weight and balance sheet. Mbuji-Mayi was so remote; everything including food had to be imported into the city. The roads were virtually useless so most of the goods had to be flown in by air. It was evident the plane was way overloaded. People even sat in the aisles along with their animals. Believe me; we had to want to go to Mbuji-Mayi to take a chance of flying in that thing. We prayed and questioned our sanity and if we had really heard from God to go as we tied ourselves in.

The captain and copilot were white foreigners, but they flew that thing like a couple of cowboys. It made us wonder how long it had been since they had taken a check ride. Loren thought they were adventurers flying in the Congo and if they died, they died. As the plane rolled down the runway we held hands and prayed. Trust me, it was not a ritualistic prayer. That was the longest takeoff roll we had ever experienced. Finally, after what felt like forever, the plane slowly lifted off the runway, but instead of rapidly gaining altitude, it leveled off about two hundred feet above the ground. We were so heavy, the pilot tried to gain enough airspeed so the plane would climb but it wouldn't climb. This didn't help our nerves. We were so low you could see people walking along the dirt road clearly. You could even tell if it was a man or a woman. It must have been at least five minutes before the plane started ever so slowly to climb. If there had been a hill in front of us, we would have never made it. Loren never said anything to me about it, but I knew what he was thinking and I silently prayed. Two hours later, we landed in Mbuji-Mayi and everyone on the plane, including us, clapped and cheered. You would have thought we had won a football game. It's no small thing to arrive safely anywhere in the Congo flying in a situation like that.

The train that was to bring our truck and equipment was delayed even longer. In desperation, we decided to fly our P.A. system, posters, and handbills otherwise the crusade would be a wash out. Before we could get that arranged, the locomotive hooked onto the rail car and started the long trip through the jungle to Mbuji-Mayi. Our team had left on a passenger train ahead of the freight train carrying our truck and equipment.

When we say, "passenger train," don't get any idea like it's anything you've seen in America. It wasn't Amtrak. The train was old and rickety. Although they rode first class, it was more primitive than you can imagine and their coach was the last car on the train. They were told to be careful because there were thugs along the route who often robbed and killed any passengers. The team forgot the warning; they were so happy the train was moving. As the trip rolled along, they came to a hairpin curve. After the main part of the train made the curve, the engineer put the throttle down and sped up. By the time our team's coach got to the hairpin curve, the train had picked up steam. In fact, it went so fast as they rounded that curve it tipped and one of the men nearly flew out the window. They were literally terrified.

Further along the trip, without notice, their car became disconnected from the train. Thank God the train was on level ground and not going uphill; otherwise they would have been on a freewheeling roller coaster. The engineer somehow realized their car had become disconnected and slowed the train down, which allowed their car to catch up and automatically reconnect before any harm could come to them.

When our team and equipment finally arrived, unbelievably we had to go through another customs inspection as if we had just entered the country again. It was a desperate struggle to get our truck and equipment released, but through prayer and a lot of effort we got them out. Because it was a diamond center, foreigners had to have a special visa to even get into the city; so we had two visas—one to enter DRC, Congo, and the other to enter Mbuji-Mayi.

Our advance man was a humble servant of God, but this trip made him more aggressive. It was either get tough or die.

Mbuji-Mayi was a massive sprawling slum city in the heart of the Congo. The tallest buildings there were wooden structures only two to three stories high at the most, but it was a complex city. The area was full of diamond mines with some of the richest deposits in the world. People could be seen in pits filled with water digging and sifting for diamonds like our old California gold miners used to do. The largest diamond ever found in Mbuji-Mayi at the time was seven hundred carats. It was a city of wealth for a few, but extreme poverty for the masses. The poor would keep digging in the hope that someday they could find a diamond that would get them out of their living hell. All along the streets you would see primitive kiosks of middlemen who would buy the raw diamonds from the miners and then sell them to brokers from Belgium and other countries.

The city was also famous for having the most advanced schools of witchcraft in all of Africa. Our crusade manager warned us he saw people roasting dogs for supper. Anything here was fair game to eat: dogs, monkeys, and even people for the cannibals.

THE MBUJI-MAYI CRUSADE

The first day of the crusade came and the crowds didn't grow as they normally do. Loren was aghast, again expecting the biggest crusade we had ever seen. Now, we had the smallest. We also discovered another evangelist had been there a month prior, but he and his team had done some things that were disgraceful to the Gospel, which turned the city off. This also was a great hindrance. It was difficult to get inspired because we had made such a tremendous effort and the results were so disappointing. He laid aside his notes and began to preach out of his heart to whatever crowd we had. Fortunately the city watched our behavior and that of our team and at least that reversed their opinion of the Gospel to those who had witnessed the worldly behavior of the previous evangelist's team. The whole situation was emotionally hard on us. It looked like the crusade was a massive failure. Sometimes you wonder if you are even called to preach. As we arrived at the field, I excitedly told Loren to look up. To our amazement the field was almost full. It was a great meeting that afternoon and we were all astonished.

Simultaneously, during the crusade, we held a minister's conference on "The Corruption of Christianity." There were so many unseemly things the pastors did in mixing witchcraft and carnality with the word of God. Immorality was a big problem also mixing their tribal traditions "making the Word of God of no effect through your tradition". Mark 7:13. The good news was that many saw their error and spent time repenting.

Salvation is a free gift that we must choose to accept. Sanctification takes teaching and understanding. "For this cause also, since the day we heard it, do not cease to pray for you, and to desire that ye might be filled with the knowledge of his will in all wisdom and spiritual understanding; that ye might walk worthy of the Lord unto all pleasing, being fruitful in every good work, and increasing in the knowledge of God;" Colossians 1:9-10.

We flew out of Mbuji-Mayi with our team to follow. Loren sat next to me in that old DC-6 with his head still hanging down and kind of blue. He looked at me and said, "I don't know what happened". I scolded him. After a few moments we laughed. It suddenly hit us; few preachers had ever preached to crowds of thousands. The laughter helped break his mood and we thanked the Lord for lifting the spirit of heaviness. Without a doubt there was great demonic opposition to this crusade. However, we learned a major spiritual lesson. It's true we wanted a massive harvest of souls, but then Loren became convicted of getting caught up in wanting to see numbers. It's not wrong to want to preach to a million people, but he realized it was carnal for him to feel that if he did not reach that number the meeting was a failure. With that restored mindset, from that day forth, we decided to do our best and the Lord would bring the harvest of whatever size it will be. We got our emotions and expectations back in order. This took the pressure off and he knew it was an answer to my prayers. Thank God for patience while the Lord worked on the situation. We realized before God can use anyone to his maximum potential in His kingdom, that person must die to himself and even his own personal ambition in the work of the Lord. That's what happened to us in Mbjui-Mayi, and it was a lesson we needed to go thru.

DEATH CAME KNOCKING AT THE DOOR

Our team stayed in Mbui-Mayi and waited for the train to take them and our equipment back to Lubumbashi. They worked hard in the daytime packing up our equipment and at night they slept on pallets on the floor of the church. Late in the afternoon in that simple wooden church, a stranger came running through the door with a mob chasing him screaming and waving knives and *pangas* (machetes).

One of our men was standing in the doorway when the man burst in. The crowd chasing the man was yelling in their Congolese language which our men didn't understand. Our team didn't know the man who raced into the church was a thief and had been accused of murdering someone. Our men thought at first he was a member of the church. The crowd gathered at the door and started wielding their *pangas* near our team member's head. He was so scared he became speechless. The angry mob broke in past him and found the man hiding in the church. They grabbed him and hacked him with their pangas with blood flying all over one of the team members and the team thought they would all be killed as well.

The mob chased the victim outside the church and continued cutting him fiercely. He tried to get up and run, but they caught him on the doorsteps of the church when he tried to re-enter. After killing him, they caught another man about twenty meters away whom they said was his accomplice and slaughtered him. The mob was the judge, jury, and executioner. After they finished their macabre work, they played their drums and danced all around the victims' bodies. This was Congo. Our team was stunned and waited in sheer terror but the mob left them alone. They all had a difficult time sleeping that night.

ESCAPE

After getting the report of what the team had gone through and because of the dangers they faced coming in on the train, we decided to fly them out as well. They made sure the truck and generator were loaded on the train properly for the return to Lubumbashi and then they headed happily for the airport as fast as they could. Unfortunately, when they got to the airport, the airline refused to honor their plane tickets and this of course caused them great distress. They let us know they wouldn't be coming in on their scheduled flight. In Africa, nothing is assured until it actually happens. It can try your fruit of the spirit. After a long ordeal of the team arguing with the agents, and several more delays, at the last minute they were finally allowed to get on a plane the following day. The men had been teasing Juma about being strapped in the seat with a rope. He was terrified of flying, but when it came down to a choice of flying, staying in Mbuji-Mayi or going on the train, he was the first one in the line to get on the plane.

Our men had to wait for our truck and equipment to arrive before they could leave Lubumbashi. Not surprisingly, the train did not arrive on schedule. After a long delay, they learned the train had jumped the tracks and were waiting for someone to set it back up. Then, after they did get it back on the track, another train in front of them had a terrible accident. This compounded the delay. It took almost a month for the truck and equipment to finally make it back to Lubumbashi. It is a miracle it made it at all, and that we didn't lose everything. Our men put the axles and wheels back on the truck and trailer and joyfully headed for the border at Zambia. When they got to the border, to their astonishment they discovered the people at customs on the border were on strike and no traffic would move across the border anytime soon. Africa. What else could happen?

They spent the night in an African guesthouse and the next morning at five A.M., someone knocked on the door. It was a woman who told them her seven-year-old daughter was sick and also a deaf mute. The woman told them she had a dream the night before that there were some men near the border who could help her and her daughter. Our team got up, dressed, and went to her home and prayed for the girl. God instantly healed her and she could hear and speak. God even worked miraculously during the delays.

When they were finally able to cross the Congolese border and entered into Zambia, the team sang and was full of pure joy. However, fifty kilometers into Zambia the truck blew a head gasket. Juma hitched a ride to a small town not far away and miraculously they had the exact head gasket they needed. It was the only one in town. Truly the hand of God worked with them.

After a long safari through Zambia, they entered Tanzania and traveled north in the mountains past Iringa. It rained as they came down a long winding road next to a steep cliff, and of course, there was no guard railing. The truck gained too much speed so our driver tried down shifting to slow the truck, but he couldn't get it to go into gear. Now they free rolled in neutral as the truck started steadily gaining speed and going around sharp curves. This was a seven-ton Iveco pulling a one ton trailer. One of the men looked out the window and saw the steep cliffs. They were perilously close to the edge and he thought for sure now they would die. At the last moment possible, our driver finally got the truck in the lower gear and was able to slow the truck down enough to get it safely under control. Our guys had been gone for almost six months on this difficult and dangerous trip. When they finally reached our home base in Kenya, we all had a prayer meeting and a celebration with food together and thanked the Lord in unison for His mighty deliverance. We agreed Mbuji-Mayi may have not been hell, but you could see it from there. While we ate together everyone noticed I was very subdued and didn't seem myself. The stress and responsibility of our team in such great perils time after time had taken its toll on me.

Chapter 15

Angels

A partner and a small team from the states had come over to see the latest churches we had built in the Masaai Mara. It was good for us to occasionally have company to come to see the work and stay excited about what the Lord was doing on the mission field. Loren carried some of the team with him in the plane. They were to meet the rest of us there. It was always great because you were sure to see wildlife either way and Loren was also able, from the air, to point out some of the churches we had built. They could be clearly seen because we had written CHI on the roof, an acronym for Combine Harvesting International. The Lord had given us the name early in the ministry. Farmers used a Combine to harvest large crops rapidly and God spoke to us to build a big harvesting machine for souls.

I took the rest of the team by road. We met the flight and as our vehicles headed into the bush we saw a pride of lions lounging close to the airstrip gorging on freshly killed buffalo. This was a treat to see. There were big males, many females, and so many young ones, maybe twenty in the group. After visiting some of the churches that team went back to the states and we continued making our plans to go up north to evangelize and build more churches in Turkana, northern Kenya, and southern Sudan. We had a strong plea from Sudanese Pastors to come up into Sudan and help them. It brought to our mind the scripture, "There stood a man of Macedonia, and prayed him, saying, Come over into Macedonia, and help us". Acts 16:9

PRELUDE TO A DISASTER

Our building team had gone ahead of us in our seven-ton 4x4 Fiat Iveco truck. We had bought a motorcycle and mounted a rack on the front of the lorry to carry it. We felt it was needful for running errands and for emergencies when we needed to send someone for help. Once again, the team had to pass through Pokot country to get into Turkana land and we always prayed for their safety when traveling that way. We would cross over into Sudan from there.

The Pokot were jealous over sharing us and the route was the domain of a lot of wildlife and snakes. One of the Turkana pastors had once been chased by three leopards for several kilometers while riding his motorcycle through this area. As would happen, our truck broke down right in the middle of this bush. The road was rough and hard on any vehicle. Juma tried to stock extra parts to travel with us but we never knew or could predict what might be needed. Our men didn't have the part needed for the repair, so Juma took the motorcycle and began the long trip back to town where he could find the parts. This left the team stranded out in the heat and in the middle of dangerous animals and raiders.

After securing the needed parts, Juma headed back to the stranded team but while crossing a sandy riverbed in Pokot country, the motorcycle slipped out from under him and landed on top of him. The steel from a footrest penetrated through his boot into his foot and left him in terrible pain and helpless. Since there was no one around to help him, he had to struggle to free himself. Self-pity would be suicidal. He managed to get the motorcycle up righted, but his foot bled badly. He toughed it out and eventually got the motorcycle going again and reached the truck to rescue the rest of the team. When others are relying on you, you do what you have to do no matter what condition you are in. They made it to the Turkana town of Lodwar where Juma was able to get some medical attention and the team set up camp to wait for us.

We had a struggle getting the lumber we needed for building the new churches but finally got a breakthrough. In the meantime, our team had been waiting for over three weeks at Lodwar in the terrible heat. Juma and our male cook had to take the motorcycle into town to replenish some supplies for the camp. Although inexperienced, the cook insisted on driving. There were big potholes in the tarmac road, but while dodging one of them, a goat darted out in front of them. The cook lost control and the motorcycle flipped and threw them both down hard onto the hot asphalt and rocky road. The asphalt tore the skin off their arms and legs. They were able to get to a small infirmary for treatment, but it ended up being a long time of suffering. This inhospitable climate made the healing process slow.

We were finally able to contact the team and let them know the lumber was on the way and so were we. We let them know we would fly to Lodwar in a few days and asked them to meet us at the airstrip. Later we would find out the team felt uneasy, knowing we had all been going through a great spiritual struggle. Things had already been tough and they earnestly began to pray fervently, sensing our flight was in grave danger. They began to fast and pray for us the next three days. Our lorry driver's wife was also a real prayer warrior and had dreamed that our plane crashed. She called and told the team about her dream and asked them to pray for our safety. We thank God for the mighty intercessors He has placed in the body of Christ. Months later we would hear other testimonies of how the Lord impressed others to pray for us as well during this time. James 5:16 says "Confess our faults one to another and pray for one another, that ye may be healed; the effectual fervent prayer of a righteous man availeth much".

OUR CESSNA CRASHES

Our Cessna 206 had come out of its required annual main-
tenance in Nairobi and we had only six hours on the engine.
After a big battle over getting the lumber and finally getting it
on its way to Turkana, we prepared ourselves and our building
coordinator and his wife to fly up north and start the work there.

Loren was apprehensive about taking off on the short grass
strip at the farm where we kept the plane. The altitude was over
six thousand feet here and that, along with the airstrip being
grass, having full fuel and the weight of four mature adults
could make the takeoff perilous. He decided, for safety's sake,
to find another runway that was much longer. We were given
permission to use an airstrip not far from us. The runway was
paved and long, which would make it much safer. We dropped
Loren off at the short strip to pick up the plane and then drove
over to the other farm to wait for him. He did a thorough pre-
flight on the plane, including the usual sumping of the fuel tanks
to make sure no water had gotten in the fuel through conden-
sation. He took off but noticed his RPM's were not as high as
normal, but nothing that was alarming.

Loren loved to fly and enjoyed the short flight over alone to
pick us up. We just had the seats re-upholstered with light tan
leather and he remarked how pleased he was at how good they
looked. After landing we put our light luggage in the cargo pod
and prepared for takeoff. Loren had topped off the fuel tanks,
so we had a good load. The wind was fairly strong that day
but shifted back and forth uncommonly. He decided to do a
pre-flight taxi alone while waiting for the wind to settle down
and while in that process, on the runway, the engine stopped.
He remembered at this altitude, there is a trick to starting and
keeping the engine going. After priming it well, the fuel had to
be put at full lean to start and then push in after it started firing;
then push the fuel mixture in, but it had to operate quite lean at
this altitude. The engine started right up and ran smoothly so
he was not alarmed.

After about half an hour the wind direction settled down and stabilized from one direction. Now he was comfortable for flying. We all agreed and got into the plane and prayed for the Lord to protect us and be with us on our flight.

Although Loren had made his calculations regarding the runway length, and knew it was well within the limits, he made one more dummy run up and down the runway to make sure things were right. That was not a normal thing for him to do, but with four lives on board and taking off at that high altitude, he said he wanted to take every precaution. Everything appeared fine and all the gauges were in the green. He put the throttle all the way in and accelerated down the runway. When he reached liftoff speed, he put down twenty-degree flaps to insure extra lift. The plane rose nicely.

He had the nose of the plane in a climbing attitude and set on his climbing airspeed. When we reached about seven hundred feet above the ground and before he had a chance to retract the flaps, without warning, the engine suddenly stopped. This came as a total surprise, because all the gauges were in the green. He had many training flights about what to do if he lost an engine, but nearly all of his training was for losing an engine at a fairly high altitude. Ready or not, there was only one thing for him to do: deal with it now. This is what was on his plate. When the engine stopped, immediately, the airspeed dropped to a dangerous level. The plane was on the verge of stalling. Since we were only about seven hundred feet, the natural instinctive thing to do is to pull back on the yoke to keep from falling further, but he knew he had to follow his training and not follow his impulses. He pushed the yoke down hard to put the nose of the plane down to try to get some wind under the wings to regain enough airspeed to keep from stalling and going into a graveyard spiral. Things happened so fast it gave him little time to find a safe place for a forced landing.

He looked around and tried to find a field or road to safely put the plane down, but there was nothing he saw that he liked.

There were small *shambas* (African farms) all around, fences, trees, and power lines. The only thing he could do was what he was trained to do under such an emergency—fly the plane first and try to keep enough airspeed up so we wouldn't stall which would kill us all for sure. The stall horn screamed. No one in the plane said a word. The others mainly looked around at the landscape but I knew what was going on. I had heard that stall horn before and knew something was very wrong and needed to be corrected fast. I sat behind Loren with the other lady, and I prayed for us. Loren didn't talk to anyone, but concentrated on flying the plane. I saw the ground coming up rapidly, and knew we were going down. The foreman seated next to Loren never said a word and appeared oblivious to any emergency.

He gently turned the plane more and more into the crossing wind, which he knew would reduce our groundspeed and enhance our survivability. If he turned too abruptly, he would lose lift in the wings. He later told us he felt absolutely no fear and amazingly, like the crash of our DC-3, death never crossed his mind. We were in a life and death situation, but his mind was totally clear and his emotions were calm. Truly that was the comforting work of the Holy Spirit. John 15:26.

As the plane descended rapidly we headed straight toward a small house. As a pilot Loren had been trained if you ever go down, if you can help it, don't take anyone on the ground with you. It's amazing how instantly all this training came back to him. He swerved to miss the house then came face to face with a line of trees and a power line intermingled with them. This of course was a grave threat to us. He had enough airspeed after swerving to miss the house and pulled back on the yoke. We barely cleared the trees and power lines. He said he couldn't see anything on the other side of the trees but pulled back on the yoke like he would if he was landing on an airstrip. The plane descended in a nose up attitude. It was then he told everyone we were going down.

Coming in blind, the plane suddenly slammed down to the earth. We hit on the cargo pod with the nose up and skidded across some furrows of a freshly plowed field and then the nose wheel came down. In a fraction of a second, the plane instantly flipped upside down. Loren's head slammed against the dash hard, but miraculously it didn't knock him out. We suddenly came to a stop. Although we were upside down, immediately and instinctively he reached up (because we were upside down which would be the floor where it was located) and turned the fuel lever off to try to prevent a fire, which he knew was an imminent threat.

Loren yelled out, "Is everyone okay?"

Everyone said they were okay, but we all knew we had to get out of the plane immediately because it was full of fuel and could explode into a ball of fire at any moment. This six seat Cessna 206 bush plane had only two main doors and a small cargo door in the back behind the third left seat. I was in the second row of seats behind Loren and was not able to get to the cargo door behind me. We hung upside down in our seatbelts. Loren tried to open his door, but it was stuck. The crash had obviously altered the fuselage of the plane and we were trapped. He hit it hard with his elbow, but to no avail. Again he hit it, but nothing. Asking the Lord under his breath to help him, he hit the door again with everything he had and the door popped open.

He unbuckled his seatbelt, tumbled down on his head, and rolled out of the plane. Then turned and yelled to me and the others "Get out now."

Everyone released their seat belts and fell on their heads to scramble quickly out of the plane. I carried our passports and money and hesitated a moment to gather my purse before getting out. Loren yelled, "Get away from the plane. She might blow."

Miraculously, none of us were killed or seriously injured. We got away from the plane about fifteen yards and lay down on the grass. We were all in shock.

Loren and I couldn't believe we had just had another plane crash. Not only were we upset about the crash, but also this troubled mission had been aborted. I lay on the ground next to Loren softly whimpering. My shoulder had been injured in the crash, but now I was mostly in shock. The other male passenger with us was able to walk and chased villagers away from the plane. There was little sympathy. Many of them would have stolen everything out of the plane if he had not stood guard. This is typical of any crash in Africa, vehicle or plane. Someone came over to check on us and out of defiance to the devil and trying to buoy everyone up; Loren asked us if we would be ready to continue our mission by that evening. It was 1:30 P.M., Easter Sunday, 2004. Everyone was stunned but readily agreed, although we knew that was not a realistic option.

Loren turned his head and looked at our fallen bird. We had mixed emotions about the ordeal. We were upset our plane had been torn up, but elated no one was killed or seriously injured. The three-blade propeller was badly bent in; the wings were all wrinkled up and bent; and the rear of the fuselage by the tail was twisted. Miraculously, the part of the fuselage where we were sitting was completely intact, as if we had been housed in a cocoon. As we looked around, we discovered we had come down in a small *shamba* (farm) that had been freshly plowed. The ground was wet from recent rains and was soft. The plane had come down crosswise to the furrows. The cargo pod had absorbed most of the impact and kept us from harm in the cabin. When the nose came down in the furrows it caused the plane to immediately flip over. The tail penetrated a barbed wire fence. We missed someone's house by ten meters, and a few meters ahead of us were railroad tracks. The plane was totaled, but the upside was that we were alive. It was truly the hand of Almighty God that we crashed where we did. It was such a narrow window to hit that spot, but for sure it saved all of our lives. I cannot understand why people don't believe God performs miracles anymore. He does them every day.

Security men had been watching us take off and after seeing the plane go down, rushed over to the site to help us. Using two of their vehicles, they took us to a small hospital in a nearby small town. My shoulder and arm were injured and Loren had a bad black eye. Our two companions had only slight injuries from seatbelt burns that happened on impact. I am so thankful we all wore seatbelts since we could have been much more seriously injured or killed had we not been buckled up. Everyone's nerves were frazzled. The television and newspaper reporters came and took photos of the crash site and then came to the hospital to interview us. By evening we had made headline news, and the next day the crash was in the national newspapers. The headline was "Four Escape Death". It was a miracle. Our God is so faithful.

After an in-depth investigation of the crash, they exonerated Loren from any wrongdoing. The cause of the crash was not caused by pilot error, but mechanical malfunction.

WE REFUSED TO LET THE DEVIL WIN

Our team anxiously waited for us at the airstrip in Lodwar, northern Kenya. They were worried when we never showed up and couldn't reach us by cell phone. They knew something was terribly wrong. Finally, they went to a public phone in the post office and called the foreman's family. It was only then they got the message that indeed we had gone down. They were told the plane had crashed, but we were all alive.

As I mentioned, the security men took us to a small hospital and they kept us overnight for observation. They released the other couple with no serious injuries. The x-ray machine didn't work, but they did their best to make us comfortable. Several people came to the hospital to check on us, including a Pentecostal Assembly of God missionary who had been working in the Baringo District. We did not know him, but he said after he heard about the plane crash he prayed for us. He said the Lord had given him a passage of scripture for us and he came to deliver that word. It was from Isaiah 54:10, "for the mountains shall depart and the hills be removed; but my kindness shall not depart from thee, neither shall the covenant of my peace be removed, saith the Lord that hath mercy on thee."

This comforted us. Not many missionaries in Kenya ever contacted us or came to our crusades. They generally treated us as outsiders. I guess because we moved around so much in the bush and other countries it made it difficult to cultivate relationships. We deeply appreciated that Brother coming late in the evening to minister to us in the hospital.

As we sat in the room in shock we knew we needed to let our family in the states know what had happened and dreaded making that phone call. We got on our satellite phone and, of course, they were stunned, but couldn't comprehend how bad the crash was and we didn't want to make a lot of it since we seemed to be alright. Our team had been languishing for three weeks in Lodwar and we were anxious to continue our mission, but now we had no means of transportation. The Lord moved thru our partners and immediately provided us with a good used Isuzu 4x4 Trooper to continue our work.

The plane crashed on Sunday; by Wednesday we had purchased the Trooper; and on Friday we resumed our mission and headed for Turkana by road only five days after the crash. We had to drive through a perilous area, which was dangerous and known for hijackings but the Lord safely led us through. After all God had delivered us from, our faith in Him did not waver at all. What a difference it was driving to Turkana instead of flying, but we all rejoiced we were at least alive and able to continue the mission.

The white sands of Turkana looked like the Sahara Desert. We saw camels roaming the desert in herds and were scattered all along our way. It was an intriguing land of mystique in its own way. We had an air conditioner in the Trooper, but it was so hot that in this heat it didn't help much. The land was so parched that thin Turkana would stand on the side of the road begging for water. They were so thirsty. It was heart wrenching. At long last, we arrived in Lodwar and had a great reunion with the team, although Juma and our male cook still hobbled badly from the wounds from their own accidents. Our cooks were mighty prayer warriors, and also gave us warm welcoming hugs and prepared some of their main staple, ugali, and goat stew. Ugali is ground maize, and in this form, is like very stiff grits and is eaten by hand and used to sop the stew.

During this tour we built fourteen churches deep in the interior of Turkana and saw so many come to Christ. We also fed the hungry with a truckload of maize. As we built, we put gutters on the churches and connected them to five hundred gallon water tanks believing God to send the rains that would occasionally come. Every ounce of water is so precious in this dry land. "O God, thou art my God, early will I seek thee: My soul thirsteth for thee, my flesh longeth for thee in a dry and thirsty land, where no water is". Psalm 63:1. We were bringing the water of life, the word of God. "But whosoever drinketh of the water that I shall give him shall never thirst;" John 4:13-14.

The team would go ahead of us getting the building started while we would evangelize the village and dedicate the churches. Many times the entire village would come out on dedication day. They responded to hear the Gospel and evangelizing the villages was such a joy to us. We always encouraged the elders of each village to use the church also as a nursery or pre-primary school. He had the authority to petition the government for a teacher once a building was up so the Church became the center of the village.

We slowly began to realize our nerves were still a wreck from the plane crash. That, along with the extreme heat, was a bad combination. We were much more emotionally fragile than we thought and began to know that we needed to slow down and heal up, but we were committed to finish the work here first. It was a very difficult time.

ANGRY ELDERS

At one of our building sites in a remote village in Turkana, instead of the usual warm greetings we normally got, we were met by a group of angry village elders. They asked what we thought we were doing there putting up a building. We told them we were here to dedicate their new church and to preach to them about Jesus Christ. They told us no one had given permission to build there. No building can be done in a village without the permission of the chief and elders. We were horrified. This was an embarrassing and dangerous situation.

We found out the district overseer who worked with us had not gone through the proper protocol by going through the elders first and this put us in an embarrassing and bad predicament. We humbly apologized to them and told them we would never have built without authorization, and would be willing to remove the church building. They listened carefully and after a tense discussion with us and among themselves, accepted our apology, and allowed the church to stand. The people in this area are extremely rough. In Africa you can never ignore proper protocol. We appreciated them for at least giving us a fair hearing and letting the church stand for the good of the village.

We continued with the dedication of the church and then went inside the village to visit with the people. We found an old woman sitting on a blanket outside her meager stick hut. She was literally skin and bones. She told us that some raiders had recently stolen her cows and she was starving to death. If you have a misfortune up here, you are completely alone. There is no social welfare net to catch you if you fail, and it appeared at this point her neighbors couldn't help. We stopped and visited with her and gave her a big portion of maize and prayed for her.

It was our pleasure to give maize to the people of the whole village. Later, we heard this precious soul didn't make it. At least she'd had a chance to hear the Gospel and make heaven.

We built a church outside Loccichogio near the border of Sudan. There must have been fifteen hundred who attended, so of course most of them couldn't get in the church. About five hundred packed themselves inside and that left no wiggle room. It looked like the whole village accepted Christ. Our vision was to plant churches where there are no salvation churches, wherever the village may be and no matter how deep in the interior. God was going before us and making a way.

Loccichogio has a big airstrip, which is used primarily as a base for the U.N. This is the jumping off point from Kenya for them to carry airdrop relief supplies to Darfur in The Sudan. One plane after another was taking off and landing there all day long. Although we had compassion on the people of Darfur, it was sad to see the UN flights and relief trucks passing the local Turkana villages that were in such dire need themselves and starving to death in plain sight.

SUDAN AND A BIG ANGEL

We finished the churches in Turkana and left the Pastors to start their work of teaching. They were accountable to the District Bishop and the Denomination from this point on. We headed for south Sudan. Our truck is a multipurpose machine used for carrying equipment for crusades as well as carrying building materials for churches. We customized the van-like box on the back, cut windows and installed bus seats inside it. When necessary we could carry the team as well as supplies inside.

We had to cross an area called "no man's land." This was a strip between the borders that was run by outlaws and the different warring factions. It would take us a couple of hours to get through this area. There was a lot of bush and was an ideal place to get ambushed. We hired two armed soldiers at the border to ride with us and also followed another vehicle in front of us that was full of soldiers armed with big guns. We prayed the whole trip. It was a hard and stressful several hours to get out of that area and to the next border.

We were surprised by the tough crossing procedures we faced at the Sudanese border station. In all our pre-trip investigations, no one had informed us we had to have special visas from Nairobi to cross the border into Sudan. This was all new information and had actually taken place while we were busy in Turkana. South Sudan became, in theory, its own nation. An envoy had established itself in Nairobi along with a President and the South had now become New Sudan.

Anyone supplying northern Sudan run by their Muslim adversaries, could not get through the border crossing. Our mission in the south of Sudan was finally recognized after long hours of deliberations. Fortunately we did have an invitation letter from the District Commissioner in Narus, Sudan, our destination, and we were traveling with the new Pastor for the church we were going to build so the military acquiesced and gave us the necessary documents and let us pass.

We arrived in Narus late in the afternoon and quickly set up our tented camp before darkness fell. We were given a spot inside the DC's property. He was our legal host but had to leave on government business for a few days so we were on our own. His wife, however, graciously let us get settled into our camp before nightfall. Early the next morning, our team set out to the church site which was located about a mile away from our campsite. The new Pastor who traveled with us was strong in the Lord and knew Swahili, English and the Sudanese language.

Loren and one of the team walked around the village to get a feel of the area and get acquainted with the people. It wasn't long until a man brusquely called them back. He said they had to go back and talk to the village elders. We followed them inside the compound and found a group of sober looking men sitting around in a semicircle on the grass. A stern looking, authoritative man sat at a simple desk in the yard. He asked in a serious tone, "What are you doing here?" The atmosphere felt menacing. We explained to him we were here to build a church and that the DC had invited us to do so. He and the elders questioned us for some time, and we decided to send for our host Bishop, Pastor Thomas to come and straighten this thing out. After Pastor Thomas came and interceded for us for some time, they appeared satisfied and let us go. It was a relief to make sure everyone understood why we were here and had the authority of the new government.

The next day, we walked to the church with the foreman's wife; wanting to see how far the men had gotten and to inspect the construction progress. On our way back to the campsite, we passed many people who were friendly to us. We wanted everyone to be aware of what we were doing and that we only meant good and were not a threat. The women usually wore only a single piece of cloth wrapped around them and many of the children ran buck-naked. It was extremely hot and the men wore shorts and a t-shirt, but all of them wore combat boots and had an AK-47 slung over their shoulder with hundreds of rounds of ammunition.

As we approached our campsite, a seemingly drunken man aggressively came up from behind us and walked around us as if to pass. As I said, everyone in Sudan is heavily armed and it appeared most of them were on drugs or were drunk most of the time. The air was thick with the heavy pungent drug smell. This man was obviously trying to intimidate Loren as the man in the group. There was no question he had the upper hand, it was his turf. He stepped right in front of us and turned wanting to fight. I was walking alongside Loren and carrying my Bible in front of me. As soon as he saw The Bible, it was as if a big Angel knocked him backward about five feet to the ground. He picked himself up and ran off in the other direction and left us alone after that. The Word of God is truly powerful and we knew it was the protective hand of God that had intervened for us. Psalm 91:11 says, "For He will give his angels charge over thee, to keep thee in all thy ways".

Foxholes had been dug in the ground all around our campsite. No one ever knew when fighting would breakout or when the Mig jetfighters from northern Sudan would come and bomb. It was smoky on the other side of the crude fence from our camp with a strong smell of marijuana. Cocaine, heavy drinking (homemade brew) along with smoking drugs combined with everyone being heavily armed made a deadly cocktail. The U.N. had given them relief grain to eat, but instead of eating all of it, they used much of it to make their brew. We were assigned a night watchman by the DC. He showed up with a heavy machine gun, got in one of the foxholes and settled in for the night to perform his vigil. I was surprised but calm in the Lord; staring at that machine gun, the building team could hardly sleep.

The following day we went to visit another Sudanese village not too far away. This is one of the most primitive places we have ever seen on the mission field. The women were half naked and most of the children were totally naked. Everyone had hideous scars all over their bodies and faces. These were considered Rite of passage and beauty markings for women, and Rite of passage macho markings for men. They represented different ceremonies they had gone through. We have never seen tougher looking people in our lives. There were no smiles. They looked totally hopeless and they were; they had been at war for so long. The men were all carrying big automatic weapons. The war in Sudan was due to the Islamic north trying to take complete control over southern Sudan and impose its Islamic sharia law on them. The south was fighting back and would not be controlled.

We preached the Gospel, gave out maize and handed out bags of clothes. It appeared all of them responded to accept Christ in genuine surrender and a sincere hope for a future. In the midst of all this, a very funny thing happened. A big man in combat boots was fighting to get a woman's pink dress. He won and immediately dressed himself in it. We tried to explain to him that it was for a woman, but he didn't care. Since he was heavily armed, no one made fun of him and we let him have his pink dress and it went well with his combat boots.

We told them of heaven where we all could live forever with Christ. In heaven there would be no more war, but peace and safety. The people were blessed so much and the Word of God filled them with thoughts of a new kind of peace. Peace that began right here and now in the midst of all their chaos. I think we were the most blessed being privileged to bring them the Gospel and love on them with the balm of Gilead, the soothing Word of God.

As we drove back to our camp, our guide instructed the driver where to drive slowly and carefully. He told us how dangerous it was for us to be here because the area was full of landmines. It seemed there were perils everywhere we turned. The next day we dedicated the church in Narus, New Sudan. What a marvelous experience. The church was packed out with some of the most primitive looking people in the world, but they were so precious in the sight God and in ours.

THE SANJO TRIBE

Our last commitment before we could think of going home was to the Sanjo people. We headed back to Kenya and south on the remote dirt roads through the Mara district and crossed over into the northern part of Tanzania. We had been moving almost non-stop since the plane crash in April and it was now November 2004. We tried to finish our goals and return to the states before the holidays to be with our family. The wear and tear of the plane crash and the constant stress of travel on the rough roads were taking an obvious toll on me by now.

This was another primitive area. The dirt roads were extremely rough, but the people were even rougher. The Sanjo worshipped a stone they believed had fallen from space and had built an altar to it. The people would gather to worship this stone and have secret ceremonies. As we drove into Sanjo land, we saw a burned out Masaai *manyatta* the Sanjo had recently attacked. They accused the Masaai of encroaching on their land. We were told the Sanjo and Masaai fought constantly in this area. The Sanjo warriors themselves were frighteningly fierce looking. Most of them wore a feather in their hair and black cloth *shukas*. We saw their altar built to the fallen stone on a little hill not far away, but kept driving past it on our way to the new church we had built nearby. We had been warned ahead of time by our missionary contact not to stop.

Many people had gathered for this day. The Christian women had decorated the inside of the church with their colorful lessos of Tanzanian cloths the women wore. This kind of decoration signified the dedication was a major celebration. The building was packed as the villagers came to hear about the Lord Jesus Christ and also to stare at the white strangers. They were appreciative of the church we had built them and there was an air of gaiety in the service. Afterward, the people had prepared a roasted goat and *Pilau*, a wonderful rice dish with aromatic spices. It was delicious. We celebrated lunch with the congregation on the grounds until we were all stuffed. Everything was so good and the fellowship so sweet in the midst of the spiritual darkness all around us.

In the early evening we returned to the pastor's home where we had set up our tents nearby on a hillside. This was not an ideal place to set up a camp because it wasn't level, but there was no flat ground around. The pastor had invited us to share their house, but we did not want to inconvenience them by putting them out of their bed. We set up our tents on the slope. Lions were in the area and had made a kill recently a kilometer from where we were camped, but we felt safe in the compound and had *askari* (guards) around the place.

It was a cold night, but that wasn't the biggest problem. We had an air mattress at the time and placed it where our head would be uphill, but we would wake up in the middle of the night with our legs hanging off the bed. It was so steep we kept sliding down. We spent four days in that area preaching. The native missionaries were fellow Kenyans and we had wonderful fellowship with their family and they fed us well. This couple had been pastors in Kenya before they went on the mission field to Tanzania. We had not met any other native Kenyans who had taken on such a missions challenge before except for our friends up in Pokot. It spoke to our hearts that God had raised up Africans to become missionaries to their own people.

We left the Sanjo and Tanzania with a peaceful feeling in spite of the harsh spiritual and physical conditions. We knew the churches we had built were in good hands. As we traveled back to our base camp in Kenya, we found our minds already had turned to going back home to the states. Two days later, as we pulled into the driveway of the guesthouse where we stayed I had a break down. I was in pretty bad shape emotionally. The last two nights before leaving Tanzania, I had exhibited some weak moments and thought I may have been bitten by a Tsitsi fly with signs of nervousness and sleeping sickness.

We left for the states the following week and Loren took me to see our doctor as soon as we got home. One day, not long after, I completely lost it and began uncontrollable sobbing. Loren found me bent over on my knees crying convulsively. I couldn't stop and would wake Loren up crying in the middle of the night almost hysterical. Some dear lady friends came over and prayed for me and ministered to me and helped me as best they could. My nerves had completely been depleted and the doctor diagnosed me with Post Traumatic Stress Syndrome.

The year had been extremely harsh with the plane crash, worrying about our children at home, Mbuji-Mayi and the life-threatening situations our team had faced. We'd had a personal confrontation with a pastor we were associated with and that, along with handling the photography and the accounting responsibilities and being the mother to the team had caught up with me. I couldn't handle any more. After three months at home, Loren told me we could wait awhile and not go back to Africa until I got stronger but I felt we needed to go or I might never go back again. This was the beginning of another long trial for us as I struggled to get myself pulled back together. In spite of this condition, we continued to minister to the hurting people there. We spent a lot of nights praying to get through the night but with the Lord's help and a lot of patience, I slowly recovered and got strong again.

Miracles can be instant but like the ten lepers, healing came over a period of time. "and it came to pass, that, as they went, they were cleansed." Luke 17:14. Literally I did what Jesus told the man who had an infirmity at the pool of Bethesda, "Arise," John 5:8. I got up and was healed.

Chapter 16
An Unexpected Guest

In the spring of 2005, we began to plant churches and evangelize among the coastal Swahili peoples of Kenya. The eastern coast of Kenya and all along the coast of Africa on the Indian Ocean has been a stronghold of Islam. Here we met the coastal Masaai, the Giriama, and the Orma tribes, among many others. We planted churches from Somalia to Tanzania. While working up in the north of Kenya near Somalia, we took a big *dhow*, a native sailboat, and went to Manda Island to dedicate new churches we had recently built in their area. There were absolutely no vehicles on these islands. Manda Island had been virtually a slave island where the inhabitants made a living in the rock quarries. They manually hewed the stones into building block size, which they would sell to the mainland in Lamu. They didn't have any donkeys to carry the rocks, so they carried them stacked several high on their heads down to the sea to load them, their bodies sweating profusely. They waded into the water and stacked them onto the *dhows*. They only got a little better than slave wages for their labor, enough to survive another day.

The people on Manda had absolutely nothing and were so grateful for the churches, which would also be used as schools. From the border of Somalia all the way south to the border of

Tanzania, churches went up in this region which had in the past been almost completely Islamic. The response to accept Christ in this forbidden region was nothing less than phenomenal. We would build churches in areas where a native evangelist from one of the denominations we worked with would be preaching to a handful of people under a tree. The work of an evangelist is so important and critical we can't under emphasize its importance. We thank God for these men and women who lay a foundation for eternal souls to go to heaven. The Bible mentions it in Acts 21:8 and 2 Timothy 4:5. "...Endure afflictions; do the work of an evangelist ". These Pastors were so grateful to have a shelter from the sun and the rains. Before we came, the Muslims in these areas would mock the Christians because they couldn't even afford a building of their own, and had to worship their God under a tree. When we brought our building teams in we would put up a church in a short time, having pre-fabricated much of it ahead of time. The churches, to the Muslims, were a real sign from God and they began listening to the Gospel and turning to the Lord themselves in great numbers. Muslim elders would come to where we dedicated a church and request that we come and build a church in their village and even donated the land. It sounds unbelievable, but it is absolutely the truth. It showed us that they were Muslim by birth and tradition but they could hear truth. That is the work of the Holy Spirit. The Bible says to "study to shew thyself approved unto God, a workman that needeth not to be ashamed, rightly dividing the word of truth". 2 Timothy 2:15. Being born again is just the beginning and opens our hearts and spirit to be taught and understand the word of God. Man did not create the terms or conditions. It was Jesus himself that said "ye must be born again". John 3:3-7, speaking of our spirit.

Not all areas were happy about our work. In one village in the interior our team arrived in the truck and unloaded their camping gear, but before they could even offload the church lumber, they were surrounded by a group of Muslim men with rocks and they tried to stone them. It became evident that their lives were in danger, so without gathering their camping gear, they jumped back in the truck and sped off while our truck was pummeled with stones. One Muslim man jumped in front of the truck and threw his *rungu* (African club) right at the windshield. Miraculously; it somehow swerved up and missed the windshield completely. This is the only place where our team has had to pack up and run for their lives while building. Needless to say, we planted that church in another village that was open to the gift we had for them, the gift of the Gospel. The Bible says in Matthew 10:14, "and whosoever shall not receive you, nor hear your words, when ye depart out of that house or city, shake the dust off your feet."

THE NAIROBI-DANDORA SLUMS CRUSADE

Dandora is one of the many large slums in the massive city of Nairobi, Kenya. We had been working with Nairobi pastors in conferences in between our village work and realized the slums needed to be reached just as badly. Almost no foreign evangelists ever held crusades in the slums. Dandora backs up to a large city dump where there are miles and miles of trash. The landscape is simply huge mounds of burning garbage. This is a dangerous place because it is the hangout of not only the poor rummaging for any leftover trash to improve their situation, but also of robbers, murderers, and of a ruthless gang called Mungiki. Juma volunteered to drive me in there to photograph this living area so our partners could somehow get a picture of where we worked. At a certain point, Juma and I realized it was too dangerous and we needed to turn back. While trying to turn the vehicle around in the tight lane we encountered a gang of young men. Keeping my cool, I rolled the window down and told the men who I was and that we were having a crusade in Dandora and asked them to come to the meeting. I handed them a flyer with our pictures and the information and urged them to be there. Because I didn't show any fear, this disarmed them.

I do have nerves of steel when required. They asked me why I was taking pictures and wisely told them I wanted a record of where the crusade was held. I gave them salvation tracts and they all agreed to come to the crusade. Later, Juma confessed he was afraid we were about to be attacked, "But God, who is rich in mercy, for his great love wherewith he loved us, even when we were dead in sins, hath quickened us together with Christ, (by grace ye are saved;)" Ephesians 2:4-5. This became our favorite scripture during that time.

Our team began to set up their tents and camp on the crusade ground. We had expensive equipment and part of their job was to make sure no one stole it. We had also hired armed watchmen to protect them. The first night two people were shot and killed a short distance from the field. After that, we got the team their own flat to stay in, and hired extra armed guards to watch the equipment.

The first night of the crusade we had an enormous crowd of expectant people standing in the open field. The power of God swept that place and many came to Christ. The Bible says, "The common people heard Him gladly". Mark 12:37. Night after night the multitude grew until the last day of the meeting when thousands crammed onto the field and beyond. So many were saved, healed, delivered, and baptized with the Holy Spirit. All the glory belongs to Jesus. John 12:32 says, "And I, if I be lifted up, I will draw all men unto me."

THE VOLATILE PASTORS CONFERENCE

We had planned a minister's conference starting the day after the crusade. An evangelical had been targeting Kenya and other African nations with an attempt at indoctrinating the pastors with his liberal theology in conferences and also through his bestselling book. From a biblical viewpoint, his teachings were full of heresy. He had come to transform the churches from spiritual institutions into social and humanitarian institutions. The purpose of our conference was to get the pastors to come back to Sound Bible Doctrine and to reject this "new spirituality" which is nothing but theological liberalism taking over Africa and Christianity worldwide including America. The Bible warns us: "For the time will come when they will not endure sound doctrine; but after their own lusts shall they draw to themselves teachers, having itching ears;" 2 Timothy 4:3. Much of our battle was to teach people to think clearly about what someone is actually saying.

Before the conference began on Monday, our conference coordinator informed Loren that another world famous televangelist had flown into Nairobi that morning in his private jet. As soon as he got out of his jet, he told the bishops, "Don't go to Davis' conference. He is a bad man." He had our street banners taken down from locations where anyone using them has to go thru the city of Nairobi to get and pay for a permit, which we had done. This bold act led us to believe he had been sent to deal with us and run us out. It also meant he had the financial and political ability to do it. The Pastors were confused because our crusade in Dandora was so mighty. We had never met this man personally and had no idea he even knew our ministry was alive. The Bishops were asking us why he would say such a thing.

We challenged the pastors and bishops to "earnestly contend for the faith once delivered to the saints". Jude 3. It was a straight forward plea, and this caused a big stir and controversy. Some pastors shouted out in disagreement and others walked out of the conference. We suspect some of those pastors and bishops were receiving financial support from certain 'new age" ministries in return for teaching their new doctrines, and they were not about to jeopardize their cash cows.

HIJACKED AT GUNPOINT

As soon as the conference was over, we walked out to the parking lot to get into our Trooper and leave with some team members. All of our vehicles and equipment were marked with CHI on the doors; Combine Harvesting International. At the last minute, our crusade manager asked us to come and ride with him instead and let the team meet us at the flat. He wanted to discuss the happenings of the day. We agreed and got out of the Trooper and into his vehicle looking forward to the discussion. He drove us back to our apartment and we had tea together. About an hour later, we got a shocking phone call from Juma. He told us they had escaped after being hijacked outside the church gate. He said they were at the police station filing a report. Right after we drove away, our team immediately followed us out of the church compound but we got separated by the traffic. As soon as they got outside the church gate, two thugs jumped them, one on each side of the car pointing pistols at their heads. Juma was told to move to the front passenger seat, and the other hijacker got in the back seat with the other two men. The first hijacker began to drive. They kept their revolvers pointed at them the whole time and drove around Nairobi for about thirty minutes threatening to kill them. When thugs hijack cars in Nairobi, they usually just kill everyone in the car, throw the bodies out and take the vehicle. These men appeared confused. They told our men to empty their pockets and took their cell phones and what little cash they had on them.

Eventually the thugs stopped and told the men to get out of the car. Our men were afraid they were about to be shot. We had given our team pepper spray for self-defense some time back and one of the men remembered he had it in his pocket. Figuring he would be shot, he decided he had nothing to lose. He reached in his pocket and took out his small canister of pepper spray and immediately began spraying one of the gunmen with it. This alarmed and frightened the thugs and they both took off running down the street. They could have killed our men right there and then, but thank God for the bravery of one man and for God's mighty deliverance.

There were some odd things about the hijacking. We had quite a bit of expensive electronic equipment in the back of the Trooper in open sight and some cash in a briefcase; the gunmen never touched them. Juma later told me he didn't think they were after things; he said since that was our personal vehicle and it was well marked, they were after us. He was convinced they were out to kidnap us. Thank God we all escaped and none of our team was hurt. The Lord is mighty to save. "The Lord is my rock, and my fortress, and my deliverer;" Psalm 18:2.

POISONED

We went back into the bush planning to visit the churches we had recently planted in the Baringo District. Our team set up the campsite next to Lake Baringo and we stayed in a small hotel next door. The following morning we walked over to the team camp and had morning tea with them and talked about our upcoming plans for the different villages. It was a beautiful morning and we were so happy visiting with them. We were close knit and always had fun together in the midst of our serious work. Our dog Rex was also a joy because he had a puppy's heart in a huge German Shepherd body. As we walked back to our place to get ourselves ready for the journey, I suddenly became ill. This attack came out of nowhere. That morning, I woke up feeling well, but in literally a split second I was almost incapacitated. It was as if something invisible had struck me down. Loren helped me get back to the room and went back to let the team know what had happened and to have them pray for me. One of our ladies came and stayed with me while Loren had to change the plans with the team. Since we were in a remote area, I didn't want Loren to take me to a local doctor, but as the day progressed I became increasingly worse. I was now so dizzy I couldn't even sit up. I complained of a headache, but nothing gave me any relief. That afternoon we were to preach an open-air meeting in the nearby town in an open market. Loren left me with the foreman's wife and our two intercessors. They were concerned about me and began to pray against witchcraft and sorcery that might have been used against me.

That afternoon Loren went with some of the team and preached in the open market. Homemade brew is a great problem in the villages and many drunks came to the meeting since it was right in the middle of town. He said a drunkard came right out in front of him in the market and challenged him. The man was infuriated to hear preaching. The team had to hold him back. People like to debate about the issue of wine in the Bible and even talk about how Jesus turned the water into wine. Had Jesus created "fermented" wine he would have been breaking the Mosaic Law. The wine he created was the best juice of the grape. He never would have drunk "fermented" wine because that would have invalidated him in being the perfect sacrifice before God. Fermentation is decay and Jesus is LIFE. Jesus said in John 14:6 says "I am the way, the truth and the life; no man cometh to the Father, but by me".

We are decaying in sin since the rebellion of Adam and Eve in the Garden of Eden. That is why we need a perfect Savior to save us. What we do in life is just an outcome of the decay we were born with from after the fall of man. Adam and Eve were created to resist the devil and had the authority to cast him out of the garden but were tempted and gave in to the temptation of believing that God might be unfair to them "...Yea, hath God said..."Genesis 3:1. Satan always challenges God's word to us. Thus, they fell and were now mere mortals and gave birth to mere mortals from that time forth.

By the time he got back to the room, Loren realized clearly my need to have a doctor. I was extremely ill and since we didn't have much choice, he sent Juma with the vehicle to go get the local government doctor and bring him to the hotel. About an hour later, Juma made it back. He had found the doctor all right, but to Loren's dismay, he was the drunk who had challenged him earlier at the open-air meeting. The man had sobered up some, but Loren was very concerned about letting him treat me. He felt if the doctor knew I was his wife he might do me some mischief. It was a nightmare that we were in this predicament.

The doctor began examining me. By this time he was more sober and began to realize the seriousness of the situation. He hadn't brought his medical bag with him and I was too sick to be moved to the clinic. He left with Juma to gather what he needed from his office. By now, I also realized the doctor had been drinking and didn't want him treating me at all. I continued to pray for the Lord's intervention. Since I was so ill, Loren had to make the decision. The situation looked life threatening.

In another hour, Juma returned again with the doctor and now it was after ten at night. We had a half drunken doctor in a primitive and remote area; a dim fifteen-watt light bulb and he wanted to put an IV in my vein to neutralize whatever made me so sick. The room was too dark and the doctor couldn't find my vein. After considerable time of trying, I begged to be left alone until morning when he could see in the daylight. He wouldn't be deterred, feeling the treatment couldn't wait. After trying several times, he finally found a vein in my right hand and gave me an IV. The next morning he came back early to check on me and to give me another IV. Five days and countless IV's later, I was finally able to get up a little. I wasn't completely stable, but at least could get up and move around. During this time our team leaders behaved very distant and unlike themselves in caring for us.

We continued to pray as we started our safari to Ngingyang among the Pokot, a little over an hour's drive away. The next few days we camped among the people, brought them more maize, and ministered to them. It was important for us to be near our flock and to take some time for me to recover. We were comfortable at our tent site. We planned to go from here up and over a nearby mountain to visit and check on one of the village churches in Barpello. Knowing the route was extremely rough for the vehicles, the ideal choice was to go the long way around the mountain, but our building foreman was insistent that we go the shorter but more dangerous route for time sake. We all knew the road; it was steep, rocky, and the cliffs along the side were a very real hazard. He would not be dissuaded and insisted the route was safe. Against our better judgment, we agreed.

About the time we were to leave, Loren became very ill and went inside the old schoolhouse and lay on the bare concrete floor. After some time, he told the team although he still didn't feel well, we would go on to Barpello anyway. Most of the trip over the mountain, he was sick with stomach problems and throwing up. When we finally got to Barpello he was so sick he just went inside the church and laid down on one of the benches. The foreman and his wife again seemed detached and unconcerned. This was the same behavior they exhibited when I was so ill in Baringo and it hadn't changed in the few weeks of this journey. I remember him standing outside the church looking through the open doorway just staring at Loren. Up to this point over the years we thought they were friends. We couldn't figure out what or why they were behaving this way but reflected on our past relationship and that made us begin to mistrust them. Sometime later it became apparent we had been poisoned. One thing we stood on was the scripture from Mark 16:18, "and if they drink any deadly thing, it shall not hurt them; they shall lay hands on the sick, and they shall recover." Praise God for that encouraging word.

Our lorry driver drove the big truck up through the pass. It was so steep that he had to off load the team to get enough power to make it over the top. Even then, at one point the truck reared up and came close to falling off the cliff. The jagged rocks totally ruined our tires and by the end of the trip when the truck finally made it back to our base, the rear end fell out. Working in these remote villages was costly and takes a big toll on our vehicles. When we count the cost in relation to souls that is part of the price we pay to reach these precious people.

It was on this trip that we found out from the pastors in these deep villages that they were being approached to accept grain for their villages. At first the pastors were elated and thought this was an answer to prayer but soon realized the strings attached could cost them their soul. The stipulation was that they could not say or preach in the name of Jesus. They could say God, but not Jesus Christ as the way to God. "And they called them and commanded them not to speak at all nor teach in the name of Jesus". Acts 4:18.

We told them we could not make that decision for them, they would have to search the scriptures and their own consciences before God and make that decision. Jesus said, "If you deny me, I will deny you before my father which is in Heaven". Matthew 10:33. We saw the world, even our remote world, being set up for what the Bible calls the mark of the beast; where no man could eat without accepting the terms of the one world government for survival. Although it was not extreme yet, it was definitely a pre-cursor to what was coming. "And he causeth all, both small and great, rich and poor, free and bond, to receive a mark in their right hand or in their foreheads: and that no man might buy or sell, save he that had the mark, or the name of the beast:" Revelation 13:16-17. "...If any man worship the beast and his image, and receive his mark in his forehead, or in his hand, the same shall drink of the wine of the wrath of God...and he shall be tormented with fire and brimstone in the presence of the holy angels, and in the presence of the Lamb:" Revelation 14:10. The Lord gave us promises to stand on, but we have to refuse this mark. "Henceforth, there is laid up for me a crown of righteousness, which the Lord, the righteous judge, shall give me at that day; and not to me only, but unto all them also that love his appearing". 2 Timothy 4:8.

PHYSICAL BATTLE

Now the door opened to begin building churches on the south coast of Kenya. This area is predominantly coastal jungle. It was a wonderful home for elephants and the big black and white Colobus Monkeys. The area was rugged and hot with high humidity. We found a multi-storied lodge built up on stilts and painted black to blend into the environment and the natural habitat surrounding it. The forest was just outside our windows. Monkeys would be playing right outside our room and on the balcony. There was a watering hole in back where elephants would come in to drink and you could see beautiful white-headed fish eagles diving to get the fish. From here we had access to go into the villages daily. Many times we would see these massive elephants as we went back and forth. A big problem here was that the elephants were coming in and eating the maize from the people's *shambas*, tearing down their huts, and at times killing people in the process. Elephants can be dangerous and unpredictable. The rangers were always alert.

One of the Muslim men who gave his life to Christ in one of these villages gave this testimony. He said when he was a young man; a missionary from England came over to this area to preach. This man and his friends hated the Gospel and were deep into black witchcraft. In their practice they did all kinds of magic to try to make the missionary leave but he just wouldn't go. One day this man and his friends completely stripped the missionary naked and ran him off in shame. He noted that ever since that time, elephants would constantly come into their *shambas* and eat their maize and destroy their huts. He publically confessed his sin and asked God to lift the curse off the area. What a tremendous salvation. As we continued to build churches deep in the bush, even here the Muslim elders from different villages would come to where we opened new churches and ask us to come and build a church for their village. It was an incredible testimony and a breakthrough for churches in the Islamic south coast.

During this time I began to have severe problems in my joints and it became so painful it affected my ability to walk. We couldn't believe these continuous attacks because we had gotten victories over the Post Traumatic Stress from the plane crash and the poisonings. I just had to refuse to let it stop the work and for a year I traveled in and out of a wheel chair. It seemed as soon as we got a victory in one area, we would face another great battle. This attack became almost completely debilitating. We saw many doctors, but they could do nothing to help me. It seemed like the enemy of our soul would pass around a tray of things to try and have me accept. This problem persisted but I praise God that I was able to be strong in the Lord; Jesus is still our helper today.

THE MERU CRUSADE VISITOR

We had been invited to hold a crusade and conference in Meru, a small city at the base of Mount Kenya about five hours drive out of Nairobi. A reporter from *The Wall Street Journal* emailed us and wanted to know where we were. He said he was on his way to do a story on us. We were amazed and couldn't understand why *The Wall Street Journal* would want to come all the way to Africa to do a story on us. Thru the years we had met other reporters but none on this level and I asked him how he heard about us but he was coy and just told us he had his sources. We told him he was welcome to come and Loren sent two of our team members to Nairobi to pick him up at the airport and bring him the five hour drive out to Meru where we were.

I was struggling with malaria and an amoeba in my system and was trying to recover before the crusade began but I managed to come downstairs and meet the reporter when he arrived. I welcomed him and told him I would not be joining him and Loren for dinner because of the malaria. We let him get settled into his room and then Loren went down to meet him for dinner. We stayed in the best hotel in town with rooms at twenty-five dollars a night plus meals. The reporter wasted no time, but immediately began delving deeply into Loren's personal life starting when he was a boy. Loren was not expecting this and realized he was an investigative reporter.

We were still not quite sure why *The Wall Street Journal* had sent him here, but we had no illusions that he was here to promote the Gospel or our ministry. Innocently we hoped whatever agenda he had, that he would be fair. After the first interview Loren became apprehensive. When he was younger he didn't have very good judgment in reading people and had made some critical mistakes. As a young man he had become a rising evangelist in a major denomination. Then catastrophe hit and because of the circumstances surrounding his personal situation, the judge gave him custody of his one-year-old son but the denomination he was with abandoned him. This set into motion a chain of events that threw his life into a horrific tailspin that lasted for twelve long torturous years. It also caused him to go into deep introspection about himself and organized Christianity. The consequences of this and other mistakes in judgement put his ministry in limbo for many years.

At this low point in his life, the Lord spoke to his heart to keep himself clean before Him and not to give up. He would use him to his fullest potential. Loren became like Moses living on the backside of the desert. He went through an intense fire. In retrospect, He could see God put him on His anvil and remade him into a highly tempered sword to be used in His hand. God, of course, had not caused his problems, it was his bad choices, but He made a bad thing turn to good. He was the vessel that had been broken and God had put him on the potter's wheel to remake him again. It was a long, painful, and lonely process. No one else believed it or in him, but he knew God wasn't finished with him. Loren remembered Job. He lost his family and everything, but he never lost his faith in God. In the end, Job found happiness and was blessed twice as much as he was before. The Gospel is about the redemption and restoration of man.

The call of God was still strong upon him, but he had no place to minister. Literally he had been boycotted but this did not stop him. Having no place to preach, he decided to witness by buying full page ads in newspapers and writing an article with the headlines "Jesus is Coming Back to Earth". This was published in different newspapers in the states and around the world. The article showed the signs of the second coming of Christ, which he correlated with current events. This article was published full page in several newspapers, including *The Dallas Morning News*, in a national newspaper of Mexico, India, two full pages in Honduras, El Salvador, and a half a page in the national newspapers of Egypt and Israel. The ones in Egypt and Israel were of course customized to work in those nations. How this happened is an incredible story in itself. His elder brother encouraged him by saying he probably reached more people even though he did not physically preach during that time.

It was in this period that Loren started his own business and began to recover. It was also during this time as a layman that he began to notice a lot of what was taught in Christianity was unscriptural. His love for Jesus and the Bible grew stronger by the day, but he began to observe and realize the church was in a real mess and began to analyze the situation. He would be listening to the Bible on cassette while working and by faith he got himself ready. It didn't matter what anyone else thought, he knew the day would come when God would raise him up again. At this point, he thought he would have been voted "the least likely to succeed." He knew the only thing that mattered was not about whom man chooses, but whom God chooses. There's only so far you can push a man down, and if he has any fight in him, he's going to try to get up.

Over the years Loren had gotten stronger spiritually and emotionally. He still made more mistakes trying to make things happen like Abraham and Sarah, in the flesh, and had to pay for those bad decisions as well. But in the heartache, he learned lessons about himself and never stopped believing God would keep his promise. He just stopped trying to get in God's way.

That's when he began to build a house believing "he had a future and a hope". Jeremiah 29:11. Then I came into his life and we got married. That's when I think he got his confidence back and began to completely heal up. I encouraged him to get back into the ministry. I would tell him, "Go Preach." I said, "our circumstances came from our failures, not God's; we will work for the Lord, but it's going to be quietly. Not many people are going to understand about us but our lives belong to Him and He can make them count. Let's go wherever He opens a door." The Lord had given us both the promise from Joel 2: He would "restore the years that the locust had eaten". We asked the Lord to extend our lives and recover those years.

In the earlier part of the book, I told you how we got started in the ministry and how we ended up going to Africa, but now at this point in our life, the Lord had blessed our ministry. We remembered God's promise to us. Loren always said I was the catalyst that triggered it. The truth is neither one of us ever expected it to rise to the level it has gone to. We thank God for His goodness and mercy in restoring us personally and our ministry. He is the healer of broken hearts and lives. "...he hath sent me (Jesus) to heal the brokenhearted...and to set at liberty them that are bruised..."Luke 4:18.

For sure you know I am not a prophet in saying we would work quietly, and we did; but now we found ourselves in Africa sitting across the table from a journalist of one of the most powerful newspapers in the world. We had no idea what he would do with this story, and were totally amazed they were interested in us at all. He began interviewing Loren the evening he arrived and proceeded year by year through his life. When he reached the "bad years," Loren tried to sugarcoat it discreetly and skip over but the journalist skillfully kept going back and interrogating him. He would not let him get away with skipping over a year. It was like the journalist was a prosecutor and Loren's life was on trial. He realized then he had been naïve about *The Wall Street Journal* coming here.

He had already experienced in Technicolor how unforgiving and cruel the religious world was. The ministry and our lives had resurrected, but now it seemed the world had sent its big guns to finish us off for good. Loren and I discussed what was happening and he sought my advice on how to respond. He was trying to be discrete as much as possible without causing a red flag. I had started college on a journalism scholarship and realized the dangers of being evasive with a journalist. I also told him the man was a professional and there was no way of hiding anything. I urged Loren to be straightforward with him. He was like Job—the thing that he "greatly feared had come" upon him. Job 3:25. We had never envisioned our ministry would grow so big and powerful that it would attract the major world news media. Like Jacob having to face Esau, we had to face our past under the highest public scrutiny.

These interviews went on from breakfast till late at night while the team set up for the crusade. At this point, it had rained every day right up to the day the crusade started. This made the crusade grounds muddy. Miraculously the Lord stopped the rain the day the crusade began until its close. Loren tried to prepare the journalist for the crusade by telling him we believed in miracles and the Lord would save and heal people in the meeting. Loren never claimed to be a miracle worker. We told him the Bible says He will confirm His Word with signs following and He also said, "I am the Lord that healeth thee." Exodus 15:26. We let him know he was free to interview anyone who came to testify of receiving a miracle independently from us. We don't fake miracles or testimonies.

The journalist was a Harvard graduate and had done his postgraduate work at Princeton. He said he was a Jew and did not believe in Jesus or in the Bible; he was an atheist. At this point Loren felt he had nothing to lose. He preached boldly. It is fairly normal in our crusades that the demon-possessed will manifest. We knew the journalist wouldn't understand and wondered how he would react if this happened in Meru. God must have a sense of humor, because one night early in the crusade while Loren prayed for the sick in mass, many demon-possessed manifested, people screamed and fell on the ground writhing like you read about in the Gospels and in our earlier crusades. Our field workers are instructed to carry them into the ICU tent where the intercessors pray and cast out the devils. We saw the journalist following them around into the ICU tent snapping pictures as fast as he could. Loren just kept ministering to the people in front of him.

Every morning the journalist would meet Loren for breakfast and immediately begin questioning him. This went on all day long right up to the time for the crusade and then would resume at the restaurant after the crusade until bedtime. It looked like he wrote down every word that was said. The Lord did many great miracles at the crusade. One night the crowd shouted and pointed down to the front where a crippled boy walked. We brought him up on the ramp of the platform so everyone could see him. The crowd literally exploded and praised God. The journalist snapped pictures of the boy as fast as he could. Loren called for the mother of the boy and she came to the platform and verified he had not been able to walk for three years. Afterward, the journalist took the boy, his mother, and her friend behind the platform and privately interviewed them for some time seeking to validate the miracle.

On Saturday night Loren preached a message titled, "Where Will You Spend Eternity?" He had to think more about the people he was preaching to than what the journalist thought. The last day of the crusade, the field was packed and so many came to Christ. The journalist came under the platform to talk to us after the meeting and asked how it felt to preach to such a large crowd. Loren's answer was that it is a sobering thing, because there are so many desperate people who need God and His intervention in their lives.

The next day we got word that a seven-year-old boy who had never walked before in his life had walked at the crusade the night before. His parents were pastors and brought him to the hotel the next day. The boy was too shy to stand up at the hotel, but six weeks later the parents called our pastor and said the boy was now perfectly whole. He was running and playing soccer.

After the crusade we had our usual two-day, all day minister's conference where Loren spoke on the issues and now included teaching about the new "emergent church movement" and its agenda. The journalist stayed with us all day long at the conference both days and again appeared to write down everything that was said.

FRONT PAGE OF *THE WALL STREET JOURNAL*

We all rode together back to Nairobi. There was no shortage of conversation. The journalist said he believed in evolution and Loren challenged him telling him it took great faith to believe the heavens, mankind, animals, and all the plants which are made in such a sophisticated and unique fashion could have come into being without a designer. The probabilities are infinitesimal. It was a contest to the last moment together.

In the ensuing month, he wrote us a few emails regarding our finances. He documented our income for the last five years. Any 501C-3 non-profit organization has to file certain documents, which are open to the public. We had been faithful with the ministry funds and I emailed our accountant to give the journalist any information he still needed. We had nothing to hide. Loren made a plea to him not to go into his "desert years". We had already suffered enough and appealed to him to consider our children. He was honest and said he would tell all but would be sensitive. He said perhaps the article might even help us. After all that was said and done, I frankly told Loren I didn't think there was a story here for him to write that would merit being in *The Wall Street Journal*. I was so wrong.

On the morning of April 25, 2006, we got an email telling us we were on *the front page* of the *Wall Street Journal*. He sent us a copy attached to the email. It was titled, "Second Chance", and spoke about how our lives were raised up out of ruins. He talked about the good, the bad, and the ugly, but he said, "Few American ministers had seen their ministry rise more steeply in Africa as has Mr. Davis'." The story brought up our past and although there were some positives in it, it was not all complimentary. This greatly traumatized us again and made public years of old personal hurts, which were displayed for public consumption. It tore open old wounds that we thought had healed up. It was like being stripped naked before the whole world.

We had mixed emotions. One was elation that the ministry was now so significant somehow it warranted being a lead story on the front page of *The Wall Street Journal*; but on the other hand, it was sheer horror the story pried so deeply into our personal lives exposing us and the Gospel to public judgement. We had not shared our past with most people. We are thankful for the mercy of God. We can't change the past, but we can change the future. The real story of our lives is the *redemptive* power of Christ to *restore* and use seemingly ruined lives.

We bought a *WSJ* and saw both our pictures on the front page in a featured story on the left hand column with a half page on the inside. Loren called to tell our close associates in the ministry so the story would not catch them off guard. When he called one of the churches we go to, before he could say anything, the secretary answered and when she recognized his voice, the first thing she said was, "Front page, *Wall Street Journal?*" Her husband had picked up a copy of the *WSJ* at the airport early that morning on his way out of town for a business trip. We didn't know what our partners would think or feel about this, but we had a long track record of having a good marriage and a productive ministry.

We praise God all our partners stood with us through this ordeal and it actually opened up an opportunity for us to begin to minister to others who had been hurt and humiliated by life. God was using us to bring healing to others who had emotionally broken lives. The Bible is full of stories of real people who had personal failure in common. So many heroes of the faith had things happen in their lives they wished they had never gone through, but when God brought it to their attention, they repented and were healed. If God could straighten them out and use them so mightily; we knew there was hope for us. The Bible says, "To whom much is forgiven, the same loveth much". Luke

7:47. We had both totally given our lives to serving the Lord and to bringing deliverance to the outcasts, hurt, and wounded. It took some time and a lot of courage, but this incident allowed the Lord to continue to work out the last vestiges of old wounds, un-forgiveness and bitterness in our own lives.

Considering everything, the only real reason we could think of why *The Wall Street Journal* would do a major front-page story on us was to mark us. We preached to such great masses of people along with the conferences on Bible prophecy in today's world. The fact that we brought attention to preaching and teaching the fundamentals of the faith in relation to the new spin on Christianity that was evolving; perhaps, they felt the true Gospel of Jesus Christ was impeding the global agenda for Africa. The "new gospel" being preached is psychology and humanitarianism. It is mixing every wind and doctrine of New Age into Christianity and is a major strategy to neutralize the True Gospel. In the global view, Jesus Christ cannot be the only savior in this new secularist gospel of attempting to meld all beliefs into one religion. It is possible we were being singled out publically to cut us off from funding. Only God and the editor knew their agenda.

A MAN SENT FROM GOD

Almost immediately after the story came out in the paper, we received an email from a gentleman who said he would like to meet us. He had seen the article in *The Wall Street Journal* and was captivated by the title, "Second Chance". He was inspired by how we had overcome so much adversity and personal tragedy to now become significant in bringing salvation to so many in Africa. We answered his email and agreed to meet. He had been a big game hunter in Africa and had over fifty trophies in his large living room, including full mounts of two leopards, a large male lion, a big horn sheep, and a huge grizzly bear, plus numerous other trophies both from North America and Africa. We had never seen anything like it.

We spent the entire day with him and his wife and showed them some of our work thru DVD's of our crusades and the work in African villages. He said the reason he was so deeply moved by our story about a "Second Chance", was that he had been a hard drinking and hard living man a few years back. He told us how God got a hold of him and "poleaxed" him and turned him around. He said he could identify with someone who had overcome such great odds and said he wanted to partner with us. "I have been doing big game hunting in Africa for many years, but now I want to do big game hunting in Africa for souls." Now, what the devil had apparently intended for evil with this newspaper expose, "God meant it unto good". Genesis 50:20. This was the beginning of a marvelous friendship.

UGANDA CONFERENCE

We flew back to Africa for a conference in Kampala, Uganda. Nearly two thousand Pastors and leaders attended our two-day conference. Thirteen hundred of them were pastors and evangelists. So many of the Pentecostal churches in Kampala had become corrupted and preached a different gospel; unsound doctrine. It was so unbelievable to us how Pentecost as a whole had changed so quickly. Many preachers faked miracles and faked the gifts of the Spirit. They exploited their people financially, promising them mammoth returns for their sowing of money into the church and ministries; things the Bible never Promises. They had gotten this from preachers in the West and Nigeria. One of their famous pastors had even written a book on the "thousand fold" return. One preacher used a hidden electrical device that would shock people and make them believe he had the power of God. Many major preachers from America had come to Kampala in their private jets and ravaged the poor who were so desperate, promising them they would become millionaires if they sowed a certain size seed into their ministries. The teaching went around that if you sowed to "poor ministries", you would be poor so you needed to "sow up"; that is, sow into

307

larger ministries so you could be rich like them. A major government official had spoken out against some of these preachers and the newspapers talked about how some of the Ugandan pastors got rich with their fast talking and taking advantage of poor Christians believing God for a miracle. Homosexuality had become rampant among the clergy. These were the pastors who were bringing in famous preachers from America who preached the "prosperity gospel." A certain pastor was confronted by a number of other pastors in Kampala about his lifestyle, but he refused to change and kept pastoring his big church. Things had gotten so corrupt in Uganda that they even had a homosexual ministerial association. The newspapers alleged he had been taken to court for defiling an untold number of young men.

We came to Kampala to contest the corruption in the Ugandan church and to try to bring the church back to its sound foundation of doctrine and godliness. We had plenty of security for the conference. The newspapers covered the conference and many of the pastors and their allies who were involved in this corruption were sent to disrupt the meeting. We had too many armed guards for them to try anything physical. The interpreter stuttered in translating. Loren had to correct him several times. Although we didn't speak the Ugandan language, the Holy Spirit let Loren know the interpreter played games with him and was distorting the message. Later we found out he was part of an organized opposition. Thankfully, most everyone knew some English and got the message correctly. They also understood what this man had tried to do.

After the morning session on the second day, we dismissed the conference and sent the people out to eat a lunch we had provided for them. We also left for lunch with our small team and two bodyguards. As soon as we left the building, one of the leaders of the opposition tried to take over the meeting, but our soundman turned the P.A. off. The people booed the man and shouted, "Let us hear the truth," so their plot failed because the people themselves knew the hidden truth. When we came back in the afternoon, Loren gave time for those who had questions or wished to clarify or dispute what The Bible said, but the opposition didn't have the nerve to confront him straight up publicly.

A newspaper reporter interviewing Loren asked where he thought the church in Uganda would be in ten years if the church continued on its present course. Loren was very candid. The Muslims had built a massive mosque on a hill in the center of Kampala with the Head of State of a North African country financing it. Loren said "unless Christianity gets its act cleaned up and gets freed from the wolves that are exploiting Christians and perverting the message, that the Christian church would be finished in Uganda". The following week, *The Sunday Vision*, the largest newspaper in Uganda, did a three quarter page write-up on one of his topics, "The Prosperity Message is Being Abused." It shook all of Uganda.

The opposition responded and began to intensify its attacks. They went on the radio and openly disputed the conference teachings and Loren's conclusions regarding the state of the church. This public wrangling encouraged the university students in Kampala and that began big debates over these issues on campus. We praised the Lord the issue had now been opened up and gotten before the public. The silence of the true men of God had empowered the wolves to continue ravaging the church. The teachings now had also come under the microscope because of the conference and the issue of truth was being heavily debated on university campuses. A great spiritual revolution had begun in the church of Uganda.

We went to a secluded hotel that night and then were secretly driven to the airport and flown out of the country the next day. Although this battle was over; the war for truth had begun.

WAR FOR THE SOUL OF THE UGANDAN CHURCH

The government was aware of the conference and they were given a set of our tapes.

We were invited to return to Kampala to do another conference as soon as possible. We returned six months later to hold a follow-up conference. Knowing it was a delicate situation, we flew in at night and met our contact and some of the team. The opposition had heard we were on our way and came to the airport ready to try to block us, but they missed us in the confusion. There was a visiting Head of State of a North African country at the same time, and the hotel we stayed in was taken over by his men. They rented the entire hotel and did not want anyone else staying there, however, we had already checked in and the manager acquiesced under pressure of our hosts and allowed us to continue to stay. Down in the lobby we could see men bowing and praying toward Mecca.

We had sent our conference and crusade team ahead to mobilize this conference and rented a nice theatre that held two thousand people right in the center of Kampala. The opposition tried to stop the meeting. An influential opposition pastor was sending text messages to all the other pastors warning them not to come to our conference. Then he went to the owner of the theatre, where he also had morning "prayer meetings," and told him he would rent the theatre for a whole week, and pay double if the owner would cancel our meeting. He was a Hindu man and told that pastor that was unethical and he would not do it. We had already paid him and he wasn't going to cancel us. The owner shared with us later that pastor's "prayer meeting" at the theatre was called the "BMW Club." He would show pictures of different BMW's and have the people confess which one they wanted. Then he would have them sow seeds for their miracle BMW's. It was a massive abuse of Christianity and the owner knew it.

When it was time for the conference to start, we were driven in a vehicle with armed guards. When Loren spoke, he had two armed soldiers flanking him on both sides of the platform, along with other plain clothed police spread throughout the crowd. Unbelievably, two hundred pastors showed up, and many of them were plants by the opposition. Loren spoke more boldly than the time before, and at times there would be outbursts from the opposition trying to sabotage the meeting, but Loren did not back down. The soldiers would bring things back under control. One of the most influential men in true Christianity in Uganda was in this meeting. He had been the point man for a major minister from the states unaware of that man's true agenda. We became friends and he respected Loren's teachings. He also had great influence in all the universities of Kampala. We gave him a copy of Loren's book, *The Paganization of Christianity.* Little did we know God would use this connection to start a spiritual revolution in Uganda. Our conference might have been small, but this pastor had attended the entire conference and wanted it aired on national radio throughout Uganda and he had the influence to do it. God has so many unique ways of working.

The next day they had arranged for Loren to do a radio interview during the morning drive time regarding the conference that was going on. When we arrived at the station there were two of Uganda's newspapers lying on the table with the headlines glaring, "Gadhafi says the Bible is Full of Errors". This was about five minutes before Loren went on the air. When the red light went on, and his interview started, the first question the interviewer asked was what he thought of the newspaper headlines that morning. Loren answered without hesitation, "That is not true. The Bible is not full of errors. The other religious book he referred to came three hundred years after the Bible. It uses part of the Bible and then adds and subtracts other things in it. Which do you believe the original version of the Bible or the one that has been revised?" Perhaps if he would have had time to think about it, he wouldn't have been so bold, but I'm sure the Holy Ghost was behind this. After Loren spoke out, many other preachers and Christians began to follow suit, boldly speaking out defending the Bible.

Our hosts had arranged to distribute the tapes of the conference throughout Uganda. Thugs seeking to destroy them broke into the house where they were to be kept, but fortunately they had been stored in another location.

After the conference the big opposition pastor went with some of his allies to see a high-ranking government official and attacked the teachings and us for coming to Uganda. The officials responded by telling them they knew who Loren Davis was and they had heard the teachings. They totally agreed with the teachings and have wanted this proclaimed in Uganda for a long time. "You preachers are defiling and robbing our people." He told them to get out of his office.

Chapter 17
A Phone Call From Heaven

On August 23, 2006 we were scheduled to preach a major crusade in the Kibera slums of Nairobi as well as hold a big ministers conference. Kibera is one of the largest slums in Africa and one of the many slums in Nairobi where millions of people live. It was on the opposite side of the city from Dandora where we had already held the very large crusade we told you about. We had compassion on the people living in the deplorable living conditions of any slum.

We were still in the States preparing to go back to the mission field for this meeting. One Sunday morning we got up and prepared to preach in a small church nearby. As we got dressed, Loren called me over and told me his right arm, hand and fingers suddenly had gone numb. He had no feelings in them whatsoever. I gave him an aspirin and began to pray for him. I was immediately alarmed and wanted to take him to the hospital because I knew these were not good signs. My mother had passed away with heart disease and I knew what those symptoms might mean. After about thirty minutes, the feeling came back into his hand and arm. He told me since the numbness had left that he wanted to go ahead and preach. He had this great faith ever since I had known him, to believe God, and he promised me we would go see a doctor the next

313

morning. On Monday morning I took him to see our doctor and he scheduled Loren for a medical screening three days before our departure date to Africa.

I have great faith in God's protecting and healing power and was not worried when we went for the screening. The test had barely begun when the screener became visibly alarmed. I leaned over his shoulder and watched the screen. He abruptly stopped the test and told us that one of the carotid arteries in Loren's neck was severely blocked. He told us to take the charts and go immediately back to see our doctor.

After looking at the screening results, the doctor told us Loren needed immediate surgery. It was Friday and he wasn't a surgeon and said he would need to make some calls to see where he could get us in to see a surgeon. Loren asked if we could do the surgery the next day because we had to leave for Africa on Monday and then suggested he could wait and have the surgery when we got back from this tour since we were already committed to be in Africa. We can look back on this now and see that blood and oxygen were definitely not getting to his brain. The doctor smiled knowing Loren didn't have a clue what kind of trouble he was in. He kept a straight face and told us calmly he wouldn't recommend waiting or leaving the country.

By this time, I entered the conversation, realizing as well that Loren had no idea what was going on in his body. I told the doctor we would get back with him on Monday after he had made some calls. Now the rest of the family got involved, insisting we could not return to Africa until we had dealt with this situation. I had to put my heels in the ground and told Loren flatly that we would not go back to Africa until we got this taken care of. He finally agreed. I called the airlines and cancelled our reservations, letting them know we had a medical emergency. The airline gave us one year to use the tickets and told me we were able to reschedule when we were ready to go. This was difficult for both of us because preparations were in full swing for this major crusade. Time was short and the funds had already come in for the crusade and were already spent setting it up. Loren called the crusade manager, and told him we would have to *postpone* the crusade until we took care of a medical crisis.

We were in a fix. We faced this expensive surgery without insurance. On Monday, the day we were to fly out to Africa, I picked up Loren's medical records at the doctor's office and we drove two and a half hours to a major county medical center to see if they could help us. We were disappointed when they told us we did not qualify for help since Loren was not having a stroke right then. We drove back home both dejected and perplexed about how God would handle this one.

We spent that Monday night tossing and turning and praying. We now felt the seriousness of the situation and that Loren's life was at stake. We decided to pack our bags and get in the car and search for help and we weren't coming home until we got this taken care of. We prayed and asked God to lead us. Our immediate plan was to go hospital to hospital and make inquiries. Our doctor had not been able to get us in anywhere and we felt we had to get up and believe God to lead us where to go. That might not have looked like a smart plan, but that was all we had. We were like the lepers in the Bible, "Why sit we here 'til we die?" 2 Kings 7:3. We headed for a good hospital we knew about that deals with heart issues. It was about an hour away. As we drove over, I asked Loren if he had remembered to call our friend we had met thru the Wall Street Journal story and let him know we had to postpone the trip. He was in Florida when Loren called and the man said he would pray with us.

When we got to the hospital, Loren pulled together all the "power of positive thinking" he could muster and with a big smile, approached the receptionist. He explained his situation and that he needed carotid surgery. She looked at Loren as though he was crazy, and without hesitation, gave him the short form answer: no. I know we looked crazy trying this, but we were knocking on doors looking for a miracle and walking in faith believing God to open that door. "Ask, and it shall be given you; seek, and ye shall find; knock, and it shall be opened unto you:". Matthew 7:7.

We sat down in the lobby trying to get our bearings and stared at each other. I suddenly said, "We don't need a hospital, we need to find a doctor first and he will open the door for us to a hospital." We tried to figure out what we could possibly do next to find a doctor; we didn't know any heart surgeons and it looked hopeless.

We hadn't been sitting in the lobby for more than five minutes after that short discussion when Loren's cell phone rang. The man on the other end said, "You don't know me." "I am a cardiovascular surgeon," and gave his name. He continued, "We have a mutual friend who told me about your situation. I can help you."

I heard Loren say, "But doctor, you don't understand. We don't have any insurance."

The doctor said, "Who said anything about insurance? When can you get here?"

Loren was so stunned because just a moment before we had prayed for a heart surgeon and now one we had never heard of before calls us on Loren's cell phone. Loren was confused and looked at me and asked the doctor to wait a moment. Almost in sheer disbelief he told me who the caller was and asked me what I thought. I quickly answered, "Let's go." Loren even asking me that question was proof of his lack of oxygen to the brain. Loren told him our car was already packed and we would immediately head that way. He told us his nurse would call us back with further instructions. Loren's brother was on his way over to meet us and a few minutes later we met him at a coffee shop nearby. When we told him what had just happened we all cried and praised God for this mighty miracle God had just done. What are the odds of a cardiovascular surgeon you had never heard of calling you at the moment you needed one? It was beyond any doubt, a phone call from heaven.

It was a six-hour drive to where the surgeon was located. It was three in the afternoon and we were on our way. We were so high we shouted and praised God for His supernatural intervention to help us and to save Loren's life. On the way, the nurse called with instructions for him not to eat anything that evening because he would have surgery early the next morning. She said she knew we didn't have insurance but to tell the hospital we were patients of her cardiovascular surgeon. They put him on the surgery schedule for the next morning. It was like we were in a dream, things were happening so fast.

When we arrived, we found ourselves in an area close to where one of our partner pastors lived. We called and told him of the great problem and the great miracle and we were here for surgery. He agreed to meet me at the hospital the next morning and stay with me until Loren was safely out of surgery. We decided to stay in the hotel across the street from the hospital so that I wouldn't be driving very far back and forth. It was more expensive than we would have liked, but we needed the convenience at this time.

Early the next morning, we checked into the hospital. It was Wednesday. The admitting clerk asked all the usual questions regarding form of payment, insurance, and so on. I told the admitting clerk we didn't have insurance and she looked down at me over her glasses and said, "You *do* know how expensive this surgery is don't you? How are you going to pay for it?"

I never blinked an eye, because of all God had done. I calmly told her it would be taken care of. That was my faith in God's provision talking. I had the foresight to bring the paperwork from the screening and also our personal income tax returns along with other things to show our financial state for the last year. They made copies and asked me for a small deposit to get Loren admitted. At seven A.M., they rolled him in to be prepped for surgery. They told me the surgery itself on the carotid should take about half an hour. Our Pastor friend came as he said he would and waited with me in the family waiting room. But, instead of half an hour, Loren was in the operating theatre for what was stretching into hours. It comforted me to have a friend with me. during the wait, but I was concerned about the length of time it was taking.

When the surgery was finally finished, the surgeon came out and introduced himself to me. I had not met him until this moment and I could see he was a calm, humble, and professional man. I was immediately impressed that he was competent and caring and instantly trusted him. He told me he wasn't able to deal with the carotid artery because when he got inside and checked Loren's heart first, he found some problems. He told me a lot of doctors don't do that, but that could prove dangerous if there were problems with the heart. He explained to me that he found that Loren had five blocked arteries; each of them blocked between seventy and ninety percent. He explained that he put stents in all five of them, but he had to do so much work, we would have to wait until Monday to work on the carotid artery.

I was quite shocked at the added seriousness of what he told me. The surgeon pulled out a drawing of Loren's heart and began to explain what he had done and urged me to make copies and keep them with us at all times. If anything were to happen to Loren, we would need those drawings to show another surgeon what he might be dealing with. Then he added, "I'm not going to send you a bill. I can't help you with the hospital, but you won't get a bill from me," I was stunned. We didn't even know this man. Truly, this was the hand of God Almighty. I was speechless but also worried not only at what the hospital bill would be, but concerned the hospital might not let us back in for the second surgery on Monday,

I went into the recovery room and saw that Loren was comfortable, and then went down to visit the hospital administrator to see if we qualified for any programs and to find out exactly what our bill was to this point. I was concerned about the finances and wanted them to know Loren was going to be dismissed on Friday but had to come back on Monday for the carotid surgery.

I wanted to be sure they would allow him back in the hospital. I was alone and it was a lot to deal with handling all this. It wasn't until later that Loren would realize what a strain this was on me as well. It was not only his medical status, but the strain of the hospital bills we faced seemed insurmountable. It felt like we needed more than one miracle and I had to believe for both of us that God was able. I could feel the struggle with the Post Traumatic Stress coming back and this added traumatic event, though I knew the Lord was helping us, was truly a lot to cope with.

Loren's son and his family drove down as soon as they could. Our other children sent flowers, called, or came after the surgery. Everything had happened so suddenly, no one had a chance to prepare; we all had to go with the flow and do the best we could. The Saturday following the surgery, a pastor of a large church in the area came by with his wife to visit us. He was a friend of Loren's brother and had learned of our situation through him. During our visit, he invited us to preach for him the following day in his church. Loren must have looked pretty good but having just come out of major heart surgery on Wednesday, we told him we would let him know how he felt the next morning. The next morning Loren called and told him we would be glad to preach at church. Four days after major heart surgery he was ministering again. The Lord gave him supernatural strength and his testimony touched the hearts of many people facing trials. It was actually a tremendous service and many came to Christ with their burdens. "Humble yourselves therefore under the mighty hand of God, that he may exalt you in due time: Casting all your care upon him; for he careth for you". 1 Peter 5:6-7. Some of the congregation were in the medical profession and were literally astonished at how strong he was. Without a doubt, his strength came from the Lord.

The following day, we returned to the hospital and they rolled Loren back into the operating theatre for a second surgery to take care of the carotid artery. A colleague of the surgeon did this surgery. They couldn't give him any anesthesia. He had to be fully alert and cooperate with the surgeon or he could die. It was a dangerous operation. The surgeon instructed Loren that when he said, "Don't move don't breathe, don't swallow," or it could be fatal, He knew he *must* not move, not breathe, and not swallow. In his mind he said, "O God, help me not to kill myself." He told me he re-consecrated his life to the Lord and repented of everything known and unknown he had ever done. Thank God, both surgeries were successful. The first surgeon told us he wanted to see Loren again in five weeks, and if everything looked okay we could return to Africa.

As they rolled him into recovery, he asked me to rebook our flight to Africa to depart five weeks from now. I was still in a lot of stress myself and didn't want to hear that just yet. I wanted to wait until he had gotten his final release from the surgeon in five weeks, but Loren insisted, knowing he would be fine. I guess after surviving all he had gone through, I didn't want him to have a heart attack over an argument about airline reservations, so I rebooked the tickets. Loren called the crusade manager in Africa and told him to reschedule Kibera. This would only throw us five weeks off schedule. James 2:17 says, "Even so, faith if it hath not works is dead, being alone."

We came to know that this medical connection came as another blessing out of *The Wall Street Journal story*. The gentleman we met thru that story had been a patient of this surgeon and when Loren called and told him we had to postpone our trip to Africa because of this medical emergency, he immediately went to prayer. He said as he prayed, he remembered the surgeon and called and told him that he had a missionary friend from Africa who needed help. The gentleman never asked the doctor to do anything except maybe just call us and give us some advice or a referral. It was he who gave the surgeon Loren's cell phone number. Sometime between that man's call to the surgeon and the surgeon's call to us, God began to deal with the surgeon to help us himself. Instead of *The Wall Street Journal* article damaging us, God used it to literally save Loren's life.

THE KIBERA SLUMS CRUSADE

At last, after five weeks, Loren was given the clearance from the surgeon and we left for Africa. The friend, whom God had used so mightily to connect us to the surgeon, flew over a couple of weeks later to meet us. He wanted to see firsthand what we were doing and how we operated. He said he was now big game hunting for souls and was excited about it. We rented a flat not far from the Kibera slums, which had good security and brought our cooks in to take care of us. Loren was now on a strict diet.

Kibera is the largest slum in Nairobi and the number two Slum in Africa. Desperate people from the villages immigrate into Nairobi and many of them end up in Kibera. They plan to stay temporarily while looking for work and trying to make a better life for themselves. The result is that hundreds of thousands of people are crammed together in a small area. They live in primitive makeshift structures made of whatever they can find. The houses share a common wall with their neighbors and the sewage runs in small trenches dug in the narrow streets and passageways. Crime and tension is high. Many different tribes are mixed together here, but they congregate in different areas. It is said in normal times, at least one person is killed every day in Kibera, but on many days there are more. Kibera has one entire section that is strictly Islamic. It is a dangerous place filled with many types of people but also filled with gangs. Not anyone can walk through and come out unscathed. Many of the people of Kibera go out into the city during the day looking for some kind of work or to beg—anything to get something to eat for that day. One friend living there told us if you have a mattress or any other belongings, you couldn't leave it while you are gone because it wouldn't be there when you got back. Most people have very few possessions.

While Loren prepared for the crusade, I had been invited by one of the pastors to come to Kibera and preach in his church. I really love being with the people and took one of our team members with me. The car could not go into the narrow passageways, so we had to get out and walk deep into those slums to bring the Gospel. Loren would praise me and tell me I had never been short on courage but I just wanted to see people changed by the love of Jesus. When I talked to him about it later, he said he never saw my face lite up with such joy. I did love this experience as much as anything I had ever done. The children walked along with me, holding my hand, and helped me step over the raw sewage. I greeted people in Swahili and invited them to come and go with me to the church to hear the good news. So many curious people seeing me walking the narrow streets in this place followed me. I'm sure my calm and assured demeanor must have been disarming to any thugs. I deliberately left the cameras at home so as not to be perceived as a "slum tourist." The photos I captured were in my mind. The people were hungry for God's Word and I went there to give them that gift.

We set up for the crusade in a big field, which backed up to the slums. We wanted to reach out to these poor people who lived wretched lives in poverty, crime, and disease. It's hard to find words to express what we witnessed. Our crusade in the Kibera slums of Nairobi was one of the most powerful crusades we have ever seen. The police said this was the largest crowd to ever gather in Kibera, even for political rallies. Jesus Christ was their most assured hope.

The power of God was so strong; it attracted many international news agencies to the crusade including journalists from England, France, and Germany. A journalist from CNN interviewed Loren from behind the platform before he went up to preach on the last day.

One of the questions he asked was, "Do you think you are doing any good in Africa?" We knew exactly where he came from with this question.

Loren answered, "We have built many churches in areas where there have never been churches before. This is deep in the interior of Africa and among many different tribes. Before we came into some areas they were raiding and killing each other.

Government officials in certain areas have told us that since we came with the Gospel, the raiding had subsided. They attributed this directly to the churches we built and the work we have done among the people in giving them hope."

He continued, "Before we started working in those areas, they were selling the girls as child brides, some of them as young as ten years old, in exchange for cattle. We have been told that this practice is subsiding. Before we came, they were forcing female circumcision on young girls. Again, that is changing." Loren explained that we not only had preached the Gospel to these people and built churches, but the church buildings were now also being used as pre-schools. Now they have many schools in the interior where there were none before. We had also brought tons of food to people in famine stricken areas as well as big water tanks which we hooked up to gutters on the church to catch rainwater.

He filmed the crusade and aired some of our story on "Inside Africa" which was aired on December 23, 2006. Later we saw *Yahoo! News* had done a three-page feature story on the crusade headlined, "Conservative Christianity Floods Africa". That article was picked up and also featured by World Wide Religious News and Medill Global Journalism out of Northwestern University.

At the end of the Kibera crusade, we had a powerful three-day minister's conference, which drew more than four hundred pastors and bishops where Loren spoke four hours a day. Despite having had two major heart surgeries in August, He was strong the whole time and got stronger every day. Our friend from the states told us he heard more Gospel in the two weeks he had been with us than he had heard the whole previous year in church. He was excited about what he had seen and heard.

Chapter 18
Crocodiles, Islam and Christmas

After the Kibera Slums meeting, we drove back to the south coast to build more churches. Our big game hunting American friend was still with us so we traveled a little slower and stopped over at a Camp for two nights to rest and recoup from the crusade. This is the actual place where one hundred and twenty five people had been eaten by man-eating lions when the railroad was built from Mombasa to Nairobi between1895-1899. The Book and movie "The Ghosts and the Darkness" was based on this true life happening in what is now the Tsavo East National Park. We were camped beside a river infested with crocodiles. At night the hippo would graze around our tents and on the trail going to breakfast in the mornings we would see their tracks where they had walked around the night before. It was interesting knowing the history of this area, but also a little eerie. There was no fence around the camp so any lion or critter could walk right in. We took our Land Rover and some time off to take our guest on a small safari. Not far from where the original camp was, where the lions had historically eaten so many, we got seriously stuck all the way up to the axle of our vehicle. It was a hot and stressful time as our men struggled trying to get the vehicle out. We were a bit isolated and it looked hopeless. We needed more man power for sure. Then from out of nowhere a truck showed up and gave us a ride to a safe place. Then they went back to help the team pull our vehicle out of the mud. In all our years in Africa, this is the only place I ever felt I didn't care to come back to. It still felt like death all around us.

MALINDI, KENYA

We had been invited to come to the coastal city of Malindi, situated on the north eastern coast of Kenya on the Indian Ocean. There was a mosque every kilometer in each direction. Men in their white robes and white fez caps and women in their black Burka's filled the streets. Many of the Burka's completely covered the woman's face except for a small slit across the eye area. There were many immigrants from Somalia as well in this city. There had been great persecution against local Christians and one of the early pastors of the city, now an elderly gentleman, told us he had been beaten by the Muslims many times for preaching the Gospel and yet he continued undaunted. He was so happy we had come to help them.

We were slightly familiar with Malindi because we had stayed in the town several times while we built churches up the coast near Somalia. Malindi was the last place that had any suitable guest house fit for us to stay in. I was also more careful with Loren now because he was on heart medication to thin his blood. I didn't want to take a chance on being too remote in case he got cut and could easily bleed to death. A key pastor, the head of one of the two ministerial associations, had been emailing us for about five years trying to get us to come for a crusade. Knowing it was a hardened city did not make Malindi an inviting target. Further down the coast, as I mentioned earlier, the Muslims had blown up a tourist hotel killing many and shot a stinger missile at an Israeli jet. Now America was at war in Iraq and the Muslims in Kenya were sympathetic to Iraq and Afghanistan. The fact that we were Americans, Christians and preachers of the Lord Jesus Christ made us prime targets for terrorists. For a long time, we never answered the pastor, but he was persistent and kept asking us to come. We only wanted the will of God to direct our paths. "Teach me thy way, O Lord and lead me in a plain path, because of mine enemies". Psalm 27:11.

Eventually God moved on this prayer and we agreed to meet with the pastors in town to discuss the possibility of a crusade. Normally, Loren and I don't attend a preliminary ministers meeting for a crusade. Our crusade manager handles these initial meetings, but this was different. Knowing how dangerous this place was, we decided to drive up from the south coast and attend the meeting ourselves. Factoring everything in, we felt in the natural, we had to have many pastors at that meeting or we wouldn't even consider coming for a crusade. It would take strong backing from the local ministries to have a chance of moving the city.

Malindi was also a tourist town and many Italians had emigrated there from Sicily. It was also filled with all sorts of witchcraft practices. On the drive to the ministers meeting, Loren told those with us in the car that he was stirred in his spirit and was reminded of Nineveh. "Arise, go to Nineveh, that great city and cry against it; for their wickedness is come up before me." Jonah 1:2.

The attendance for the meeting was frankly disappointing. It was about half the number we had set in our mind for what we felt was necessary to have a successful crusade. Loren sat on the front row, skeptically, with his arms folded and listened as the meeting began.

The pastor began to speak. He explained to us that Satan had bound the city for as long as it had existed. He told us how a covenant had been made with Sicily where people could freely travel back and forth to do business or intermarry without interference. The tourist influence was strong here and the locals were deeply involved with them in gambling and prostitution. The pastor told us during the heavy tourist season, many of Malindi's husbands and wives would separate and go into prostitution until tourist season ended, then get back together and live as husband and wife. He made a great plea for us to come to the city and bring deliverance from the bondage of Satan. As he spoke, I saw Loren's arms begin to unfold and tears flowed down his cheeks. I had also been concerned about coming here, but as the pastor spoke, God touched my heart as well. We had never heard such an impassioned plea for a city. Now we knew we would come even if only one pastor was behind the meeting. God had compassion on Malindi the way He had about Nineveh. "Arise, go unto Nineveh, that great city, and preach". Jonah 3:2.

Though this was a strongly Islamic town, we did not come in passively. We came in like a storm. The Bible says, "The righteous are bold as a lion". Proverbs 28:1. We had ten full color street banners, five thousand full color posters, and two hundred thousand handbills. A close pastor friend of ours and his assistant pastor flew in from the States to be a part of that crusade. Before the meeting ever got started, the demons of hell broke loose. We have learned when God is getting ready to do something big is when the greatest opposition arises. The Bible says regarding Satan, "He knoweth he has but a short time". Revelation 12:12.

HUMAN SACRIFICES

It was discovered that Islamic elders had made blood sacrifices to try to stop the crusade from taking place. One day twenty aborted babies were found under a ceremonial tree in town. They had been sacrificed as an offering to Satan to bring a curse on the crusade and stop it. I was heartbroken when I heard of this, and reminded the team and the Pastors that there were twenty women who were also suffering from being forced to abort their children to their "god." Although we didn't know these women, we prayed for them. We asked God to open their spiritual eyes in this meeting so they could see. 2 Corinthians 4:4 says, "The god of this world hath blinded the minds of them which believe not, lest the light of the glorious gospel of Christ who is the image of God, should shine unto them."

Some of the pastors told us one evening while they walked by the ocean praying for the city and the crusade, they saw a large ghostly looking woman rise up out of the sea. They said it rose up and floated in the air above the water and lingered there for a time. Many Africans believe demons live in the ocean. The Muslims and witches took this as a sign their prayer was heard and the meeting would be stopped.

As if the Muslim opposition and witchcraft wasn't enough, another major ministerial association in town united against us. We had seen this so many times before and always made a plea for unity. When we were informed about it, we asked our advance pastor to approach them about joining with us and working together. They told him bluntly, "Not unless you come to town through us and give us money." We were not going to be blackmailed, and continued to work with the pastors who had invited us. It was amazing that we continued to run into this tactic of controlling the work of God for money no matter where we went.

Some of the opposition pastors went to the District Commissioner and petitioned him to stop the crusade. The DC was Muslim, but incredibly, he told them no, he would not stop the crusade. He told them that witchdoctors terrorized and ruled the villages and the police couldn't handle the situation. He told them he had already given Davis a permit to hold the crusade. He said, "Let Davis do the crusade. Let's see what he can do to help the city."

The opposing group continued to try to stop us from holding our meeting. Some of the opposing pastors of the large churches organized a huge march protesting the crusade. They went through the city telling everyone not to come. It was amazing that our biggest opposition was not from the Muslim community, but from "Christian" pastors. However, people who attended their churches paid no attention to them.

Pastors who backed us organized a march and went throughout the city as well. It became a parade. Even people from the opposition churches joined the crowd in walking thru the streets singing and handing out handbills about the meeting. Our friends from the states joined in the parade. They had come over to support us in this hardened place and were working in every way to ensure success. Christian music blared; everyone sang; and our big Lorry (truck) followed with our billboard sized banners advertising the crusade on the truck sides. It was such a bold move for the Christian community to do such a thing. The parade couldn't help going past the mosques because they were everywhere. The elders were sitting on the porches just staring at the commotion as it went by.

The team began to set up our big equipment on the field right in the heart of Malindi. Muslims wanted to know why we had so many armed guards and the team told them it was because our equipment was expensive. The truth was they were there to make sure there was no violence against the meeting. We had no idea what to expect out of the people of Malindi and the police wanted to be sure there was no incident. We were both assigned armed bodyguards for security and came to the crusade in an anonymous car every day and via a different route.

People asked Loren what he was going to preach in that Muslim city. He told them he would preach Jesus Christ, the only begotten Son of God; that He was crucified for our transgressions (sin), God accepting Him as the perfect, sinless substitute in our place. Then God proved the sacrifice was acceptable and raised Jesus from the dead to show this acceptance; He is the way, the truth, and the life, the only way to heaven. He is not dead, but alive and still does miracles today. He would preach the straight undiluted Gospel. If we weren't going to do that, why come and hazard our lives? "And they overcame him (satan) by the blood of the Lamb (Jesus Christ), and by the word of their testimony (they told others); and they loved not their lives unto the death". Revelation 12:11.

We had a good attendance the first night and Loren boldly preached Jesus. There were no protests. Many responded to accept Christ and then the Lord began to do many signs and wonders with great miracles in the name of His Son Jesus Christ. Jesus himself tells in his word, "Except ye see signs and wonders, ye will not believe". John 4:48. "...for unclean spirits, crying with loud voice, came out of many that were possessed with them: and many taken with palsies, and that were lame, were healed. And there was great joy in that city". Acts 8:6-8.

Another night Loren preached about the power of the blood of Jesus. A former Pentecostal Bishop was there who had backslid and converted to Islam in exchange for a large sum of money. As Loren preached, we were told this man came up close to the crowd control rope and stared intently at him as he spoke so boldly. The man listened carefully to every word that was said. He was observed by local believers who told our team who he was. After hearing the message about what made Jesus different from a normal man; that He had God's blood in a man's body, our team told us he went back and began to talk to fifteen other Muslim elders in the back of the field. He was overheard to say "This man is telling the truth about Jesus." Muslims do not believe Allah, their god, had a Son. They accept only that Jesus was a prophet. Muslims do not believe Jesus died on the cross; they believe an imposter died in His place. Now, that backslider had seen Jesus afresh and tried to convince his Muslim friends of the truth. Jesus was God. "I and my father are one". John 10:30.

Many cripples were healed in the services. It was amazing. Loren didn't lay hands on anyone, but at the end of the services every night after the call to accept Christ as their personal Savior, he prayed for the multitude as a mass. Faith built higher and higher every night as the Lord healed so many people. We began to notice people spontaneously brought chairs in front of the crowd control rope facing the platform and put their crippled loved ones in them. Many of those were healed and got up and walked, testified, danced, shouted, and praised God. A mama and daddy brought their six-year-old son, holding him by each arm and dragging him. His impotent legs dangled behind as they brought him to the front. They sat him in a chair there on the front row. Loren was preaching about the resurrection power of Christ and led people in a prayer of repentance and prayed for the sick. Suddenly, the boy stood up out of his chair. Some of our team working in that area were watching him. They told us the mother grabbed the boy by the shoulders and sat him back down. This happened several times. Seeing what was happening, one of our men realized God had healed the boy and the mother didn't know what had happened. He went over to her and said, "Mama, the Lord has healed him. Let him get up and walk." This time she left him alone and the boy got up and walked. Seeing the miracles, Muslims began to believe that Jesus truly is the Son of God and accepted Him as their Savior in great numbers. Our pastors were getting excited with new faith arising in them.

One lady in the audience had been abandoned and cast out of her home by her husband. She lifted her hands to give her life to Jesus and repented of her sins. Then she came to the front for prayer. As she prayed, she told us her mobile phone rang. It was her husband. Right then and there, he apologized for how he had treated her by casting her out of the house. He told her he loved her and asked her to come home. This is not the behavior of the typical African male. Our team member watched her when she got the call and noticed a great change and joy come over her. She came to the platform and told everyone what Jesus had done for her, saving her soul and restoring her broken home that night.

The field was packed with people crying out to Jesus in great faith to save and heal them. That night a little girl, seven or eight years old, who had never walked in her life, was carried to the meeting on her big sister's back. The little girl got up and walked, miraculously healed by the power of God. She and her sister came to the platform and told what the Lord had done for her and everyone watched the little girl walk for the first time in her life. Physical miracles are an outward sign of what God can do on the inside of a person. If He can take care of the outside, how much more is He able to change you on the inside? You have never heard a crowd roar louder or seen greater joy than we witnessed that night. People were so blessed and realized only Jesus Christ can do such a wonderful thing. We can only imagine what the eyes of the parents were like that night when their little girl came walking home.

Christians from all over Kenya called and asked, "What's happening in Malindi?" The news had spread. We did not know until the next day that television reporters from Kenya's national television had come down to cover the crusade from Nairobi. They filmed the crusade the last night. The message was about how it was impossible for an imposter to have died on the cross. Loren's argument was that it was the Roman Government who crucified Jesus and they could not be fooled. They knew it was Jesus. Everyone was at His crucifixion. People had seen Him and heard Him preach many times. It was one of His own disciples, Judas, who identified Him and betrayed Him. Jesus mother was at the crucifixion. He was stripped nearly naked. The Bible says in Isaiah 52:14, "His visage was marred," but Mary knew it was her son Jesus on that cross. Loren argued, "You can fool a lot of people, but you can't fool a mama."

After the meeting that night, Muslim women were interviewed by KBC television at the back of the field. We learned about it when it aired the next day. The Muslim women said "We have been deceived about Jesus. You can't fool a mama. It was Jesus who was crucified and rose again the third day. We now believe in Him. He is more than a prophet".

"For whosoever shall call upon the name of the Lord shall be saved. How then shall they call on him in whom they have not believed? And how shall they believe on him of whom they have not heard? And how shall they hear without a preacher? And how shall they preach except they be sent"? Romans 10:13-15a.

A PERSONAL GIFT

We went back to the states with a wonderful sense of gratitude for all the Lord had done and excited to be spending Christmas with our family. We continued to work to complete our house and it was not far from being finished, but we still lived on wall-to-wall concrete floors. We often said we went to Africa for relief. Trying to work on the house and keep going to Africa for five to six months or more a year , may have seemed like a strain to some people but we put the Kingdom of God first. One day we got a phone call from a partner saying they wanted to give us a personal gift. We received a significant check, enough to buy nice carpet for our home. It was a real testimony to our family. We praised God for this personal miracle and it would make our future homecomings much more comfortable and enjoyable.

One thing clouded our thoughts; the hospital and surgery bills. In the months we were overseas, we expected to come back to find the outstanding balance due. Surprisingly it wasn't in the mail. We were just about to call about it when two days before Christmas the phone rang. It was the hospital business office wanting to speak to us about the bills. We both got on the line and the voice on the phone said, "Mr. and Mrs. Davis are you sitting down?" We looked at each other in anticipation of the shock; then she said, "The Hospital has decided to forgive your bill, Merry Christmas!" Almost speechless, we stuttered out "Thank you, and please give our thanks to the hospital and a Blessed Christmas to all of you"!

We hung up the phone giving a prayer of thanks to our wonderful Lord for the Doctors and the Hospital he had in place to help us during that time of personal crisis. Then, we remembered a promise from His word. "But my God shall supply all your needs according to His riches in glory by Christ Jesus". Philippians 4:19.

"Goodness and mercy *shall* follow me all the days of my life"…Psalms 23:6.

Afterword

The overreaching goal of Mission's work can happen by simply following where the Lord opens a door. For us, it was as in the book of ACTS. There was no "master plan" devised by us, only a burning desire to serve the Lord and to give our lives to something bigger than ourselves.

We dove in with our whole heart, forsaking any "back-up" plan if it didn't work out. Failure to obey Him was not an option. We trusted the Spirit of God and the Word of God to lead us.

We continued to work in Africa in crusades and conferences until November of 2014.

With the help of our partners we built a total of two hundred and seventy six churches in unreached villages of East Africa, most of which are also used as pre-schools. These churches were given to the community and overseen by the various denominations we worked with.

India opened up to us and we began crusades and conferences in that country in 1994. Those meetings were preserved on video and DVD and are aired on District Television throughout India and continue to bring people into the family of God.

World-Wide shortwave radio carries "The Great Commission Crusades of the Air" with Loren Davis on seven separate satellites reaching unreached countries around the world. We receive letters from listeners who are able to hear the gospel and receive spiritual nourishment in uninviting environments.

The ministry is on a network of One hundred and Twenty Six radio stations which airs programing provided by Combine Harvesting International Outreach Church and Loren Davis Ministries from Africa to Beijing China.

My precious husband, Loren Davis went to be with the Lord on July 20, 2015.

The rumors of his death are greatly exaggerated.

He is alive today in the Heaven he preached about but is also still preaching here on earth thru technology and the men, women and young people his preaching inspired and raised up to carry on the spreading of the Gospel of our Lord Jesus Christ.

Well done thou good and faithful servant...Matthew 25:21

I continue to share the Gospel around the world thru speaking, writing and electronic media.

The work of this ministry is 100% provided for by your donations.
Combine Harvesting International Outreach Church is an approved charitable non-profit organization.
If our story has encouraged you and you would like to learn more about our current outreach,
We encourage you to visit us at
www.lorendavis.com

CPSIA information can be obtained
at www.ICGtesting.com
Printed in the USA
LVHW111521280421
685858LV00005B/99

9 781498 474214